OXFORD LATIN COURSE

PART III

£4-50

q

MAURICE BALME AND JAMES MORWOOD

Oxford University Press 1988

Oxford University Press, Walton Street, Oxford OX2 6DP

Oxford New York Toronto
Delhi Bombay Calcutta Madras Karachi
Petaling Jaya Singapore Hong Kong Tokyo
Nairobi Dar es Salaam Cape Town
Melbourne Auckland

and associated companies in
Beirut Berlin Ibadan Nicosia

Oxford is a trade mark of Oxford University Press

Acknowledgements

The illustrations are by **Cathy Balme**.

The cover lettering is by Tony Forster.

The authors would like to express their warmest thanks to George Littlejohn of Smithycroft Secondary School, to Professor E.J. Kenney of Peterhouse, Cambridge, and to Dr. Jonathan Powell of Newcastle University for their generous and helpful advice.
They would also like to thank the University of Michigan Press for permission to reproduce extracts from *Gaius Valerius Catullus: The Complete Poetry*, translated, with an introduction, by Frank O. Copley, copyright © 1957 by the University of Michigan; and the executors of the estate of C. Day Lewis and The Hogarth Press for permission to reproduce an extract from *The Aeneid of Virgil*, translated by C. Day Lewis.

The publishers would like to thank the following for permission to reproduce photographs:

Ancient Art and Architecture Collection p. 14,131,193; Archeological Museum, Naples p. 154; Bibliothèque Nationale p. 194 left; Oxford, Bodleian Library, MS. ADD. C. 139, fol. 96^R p. 66; British Museum p. 10 top, 33 right, 89,101,102,117,119,122,132 bottom, 136,143,152,179,186,208 bottom, 220 top; J. Allan Cash p. 190,193; R.L. Dalladay p. 30, 41,43,140 bottom, 188,206,207,208 top; Mike Dixon/Photoresources p. 9,49,73,138 top, 139; Fitzwilliam Museum p. 189; Werner Foreman Archive p. 60,91; Fototeca Unione for the American Academy, Rome p. 23; Giraudon p. 224 bottom; Sonia Halliday p. 75,118, 163; Michael Holford p. 135,167; A.F. Kersting p. 217; Louvre p. 220 bottom; Mairani/ Grazia Neri p. 52,172,183,191; Mansell Collection p. 10 bottom, 12,18,50,51,65,67,76,94, 128 top, 104 bottom, 144,162,196,201; James Morwood p. 29,32,96,213,224 top; Museum of Fine Arts, Boston p. 138 bottom; National Museum of Athens p. 102; National Museum of Wales p. 20; New York University p. 21; Octopus Books Ltd. p. 130 left, 142,156; Fabrizio Parisio p. 53; Rheinisches Landesmuseum, Trier p. 192; Roger-Viollet p. 128 bottom, 165,204; Scala COVER, p. 130 right, 140 top, 151,200,202; Edwin Smith p. 70,126; Somerset County Museums p. 149; Graham Tingay p. 24,33; Vatican Library, title page, 130 left, 132 top, 160; Vatican Museums p. 48; John Wilson p. 218.

Cover photo: Scala

Set by Tradespools Ltd, Frome, Somerset
Printed in Great Britain by Butler & Tanner Ltd, Frome.

CONTENTS

INTRODUCTION

The third part of this course consists of extracts from authors who belong to what is called the Golden Age of Latin literature. This was the first flowering of literature in Rome, which lasted from about 75 BC to AD 14 (the death of Augustus), and most would agree that this period produced the greatest writers in Latin. All were contemporaries of Horace, the hero of the first two parts of this course. He knew most of them personally and was friends with several, notably Virgil, whom he describes as 'half of my soul'.

The writers fall into two groups. The first were rather older than Horace (65–8 BC) and wrote during the last years of the Republic. They were Caesar (100–44 BC), Cicero (106–43 BC), and the poet Catullus (84–54 BC). The second group are known as the Augustans, since they all wrote during the reign of the first Roman emperor, Augustus. They were Virgil (70–19 BC), Livy the historian (59 BC – AD 17) and the poet Ovid (43 BC – AD 17).

We think of Caesar as a statesman and above all a general, the conqueror of all Gaul. He was also a highly cultivated man, a brilliant speaker (second only to Cicero, according to contemporaries), a student of literature, a critic, a friend of poets, an admirer of Cicero, and himself a writer. His best known work is his *Commentaries on the War in Gaul* (*Commentāriī dē Bellō Gallicō*), in seven books, covering his campaigns from 58 to 52 BC. It is from this that you will read extracts on his two invasions of Britain (55 and 54 BC) and the revolt in Gaul which took place the following winter.

The *Commentaries* were a new form of literature. They are a kind of war diary in which Caesar gives an account of his campaigns year by year. He writes for the public in Rome to tell them what he has done and to justify the continual extension of his conquests. He writes in the third person, calling himself Caesar, and appears to give a straightforward, factual account of a patriotic Roman fighting necessary wars. Indeed, there is no reason to doubt the truth of what he says at any point, but he arranges the facts in such a way that Caesar always appears in a good light and never makes a mistake.

Nevertheless, Caesar clearly emerges as one of the world's greatest generals. He made the right decisions both strategically and tactically; he weighed up situations with absolute clarity and acted with amazing speed; he showed great personal courage; he was prepared to take calculated risks when the situation required this but did not hazard the lives of his troops unnecessarily; he had the complete confidence of his men; he was a merciless critic of those who let down him or themselves but was generous in praise to those who had done well. All this is illustrated in the passages you are going to read.

Caesar writes in a lucid and simple style which mirrors the clarity of vision which characterizes his actions as statesman and general. Cicero describes the *Commentaries* as 'bare, direct, graceful, stripped of all adornment of style'. The story moves swiftly and he seldom pauses to comment on the action. The reader is carried along by the pace of the narrative, an exciting story, which we hope you will enjoy.

If Caesar is the ideal man of action, clear-headed and quick to act on his decisions, Cicero is a very different character. He was by nature a thinker, a scholar and a literary man, who was driven by ambition to a career in the law courts and the hurly-burly of Roman politics. His literary output is phenomenal. He wrote a good deal of poetry (perhaps rather poor stuff), seven works on rhetoric and seventeen on philosophy. Fifty-eight of his speeches survive from a career during which he dominated the Roman bar. Whenever events forced him to withdraw from politics for a time, we find him writing. And in his writing he reshaped the Latin language; his influence remained incalculable throughout ancient times and right up to the last century.

His letters are perhaps the most interesting and attractive of all his works. Thirty-five books of his letters survive: fifteen books to his friend Atticus, fifteen to his family and other friends, collected and published by his secretary Tiro after his death, and three books to his brother Quintus. The letters vary from quick notes dashed off in a hurry to long discussions on the political situation or literary subjects. They form a unique collection, which throws an amazing light both on the politics of the time and on Cicero's own character. From them we get to know him better than any other man of ancient times. Besides Cicero's own letters, the collection includes replies from many of the leading figures of the time, including Julius Caesar.

Chapter 5 gives an outline of Cicero's earlier life and to illustrate this we have used extracts from dialogues and speeches. You are likely to find the passages from the speeches difficult compared with Caesar. Cicero's style of public writing is more complex and less direct than Caesar's. He uses the adornments of rhetoric which he finds missing in Caesar. For instance, he likes to use long sentences with lists of parallel clauses or phrases, rising to a climax. You should always remember that he wrote his speeches to be heard by an audience whose emotions he intended to sway. Many of his stylistic tricks are still used by public speakers today. You should read these passages aloud; their rhythms help to point the meaning and what looks complicated on paper often becomes clearer when it is heard as it was intended.

You have already read letters of Cicero to his brother Quintus, to Caesar and to Trebatius, his lawyer friend. The letters are mostly written, as you would expect, in a much freer and easier style than the speeches; apart from colloquialisms (which we always gloss), they are not hard. We include extracts from five letters to Atticus, three to his

wife, Terentia, another to his brother Quintus, letters to and from
Caelius, a young friend who was keeping Cicero in touch with affairs
in Rome, while he was away governing the distant province of Cilicia;
a letter from Caesar written at the beginning of the Civil War, in
which he tactfully presses Cicero to join his side; one from an old
friend, Servius Sulpicius, offering condolences on the death of
Cicero's daughter Tullia; and one from another friend, Trebonius,
who on his way through Athens had met the young Marcus Cicero and
wrote giving Cicero a good report on his son.

A clear picture of Cicero's character emerges from these letters.
He was often anxious and self-critical; vain but well able to laugh at
himself; filled with self-doubts but in the last resort high principled
and courageous; loyal and affectionate, impulsive and emotional. It is
surprising that a man of so complex and humane a character should
have survived so long and achieved so much in the last turbulent years
of the Republic.

Catullus (87–58 BC) was born in Verona in the north of Italy but
moved to Rome as a young man and soon joined a circle of young
poets who established a new tradition in Roman poetry. Up to this
time Roman poetry had been limited almost entirely to the forms of
epic and drama, both intended for public performance. It may seem
extraordinary to us, but poetry had not been used as a vehicle for
expressing personal feelings. Catullus and his circle, whom Cicero
called the **poētae novī**, broke with this tradition consciously and
completely.

At the head of the published collection of his shorter poems
Catullus puts a brief poem of dedication, which begins.

cui dōnō lepidum novum libellum. . . ?

To whom do I dedicate this elegant, new little book?
The adjectives characterize the contents as well as the appearance of
the book. The poems are witty, original and small-scale, and he writes
to express his feelings on every subject that comes to hand, from the
trivial to the profound.

He was the first Roman love poet. Chapter 10 contains a selection
of the poems which he wrote about his passionate and unhappy affair
with Clodia, the sister of Cicero's enemy Clodius, who was stolen
from him by Caelius, his friend and Cicero's.

Catullus is a poet who uses the language and rhythms of everyday
speech with splendid effect; indeed, many of his poems are studded
with colloquialisms. You will find that most of the poems seem the
natural expression of universal emotions, but beneath their apparent
spontaneity there lies a great deal of poetic art, seen most clearly in
their careful construction.

No further introduction is necessary for a poet who has appealed
to the young of all times. The poems are not difficult and we hope we
have not overburdened them with notes, being warned by W.B. Yeats:

Bald heads forgetful of their sins,
Old, learned, respectable bald heads
Edit and annotate the lines
That young men, tossing on their beds,
Rhymed out in love's despair
To flatter beauty's ignorant ear.

All shuffle there; all cough in ink;
All wear the carpet with their shoes;
All think what other people think;
All know the man their neighbour knows.
Lord, what would they say
Did their Catullus walk that way?

The greatest poet of this extraordinary flowering of literature was Publius Vergilius Maro, known in English as Virgil. He was born in Mantua in North Italy in 70 BC, and grew up amid the terrible civil wars which rent the Roman world. His early poetry celebrated the beauty of the Italian countryside but set it against the background of civil strife. This had caused a quarter of the land of Italy to change hands in the proscriptions and confiscations, and had taken the farmers from the soil and pressed them into arms. Like most of his contemporaries, Virgil was profoundly grateful when Augustus reestablished peace after the Battle of Actium in 31 BC.

In his last and finest poem, the *Aeneid*, Virgil expressed what he felt about Augustan Rome, but he chose a legend from the distant past in order to do this. The story concerns the flight of the Trojan Aeneas from the ruins of his sacked city and his journey to Italy where he establishes a foothold for the Roman nation. Virgil fully recognized the greatness and importance of Aeneas's Roman mission. It was eventually to lead to the new golden era of peace and stability created by Augustus. But he was equally aware of the suffering caused not only to their enemies but also to the Romans themselves by their imperial destiny. We have included Virgil's story of Dido, the Carthaginian queen who is tragically destroyed by her love affair with Aeneas. Virgil asks us to consider whether even the greatness of Rome is sufficient to justify such torment.

Some of the finest passages in the *Aeneid* proudly proclaim the glory of the city which civilized the Western world. But there is a more private voice to be heard. No poet has seen deeper into the mysteries of the human heart or conveyed more movingly the pain involved in human experience. The *Aeneid* was incomplete when Virgil died in 19 BC and the poet asked that it should be destroyed. Augustus overruled this request. Thus we owe it to the Emperor that we can read the only poem in Latin able to stand comparison with Homer's great epics, the *Iliad* and the *Odyssey*.

Virgil's rather younger contemporary, the historian Livy, wrote his huge history of Rome – it was in 142 books – with the intention of illustrating the qualities which had brought the nation to greatness, and then of describing what he saw as its moral decline. His comments on the value of history are interesting:

'What makes the study of history a particularly helpful medicine is that it sets before you a clear record of the infinite variety of human experience. This enables you to find models for yourself and your country to imitate, as well as courses of action which were disgraceful from start to finish and should be avoided.'

History is good for you because it educates you.

We give one of the highlights of Livy's history, his description of Hannibal's crossing of the Alps. Livy's Latin is difficult and we have cut out the most problematic passages. But he is always exciting and he has the gift of making his readers live through the events he describes. He conveys with startling immediacy the despair of the Carthaginians on the frozen heights of the Alps, and then, as Hannibal moves forward onto a ridge and addresses his troops, the historian vividly illustrates the qualities which go to make a great general. Livy responds with a wide-eyed admiration to the epic nature of Hannibal's journey and conveys the dreadful threat posed by an enemy which was to test the courage and determination of the Romans to the full.

The last of the great Augustan poets was Publius Ovidius Naso. The range of his poetry is wider than any of his predecessors', and in this book you will be sampling his autobiographical writings, his love poetry and his huge work about mythological subjects, the *Metamorphoses*.

Ovid's love poems do not aim at the passionate intensity of Catullus or the profound melancholy of Virgil. For him, being in love is fun, and even when things go wrong, it is not a matter for serious grief. A cheerful sensualist in his poetry, he offended the Emperor, who was trying to clean up the morals of Rome. He was exiled to a remote spot on the Black Sea and in these bleak surroundings his writing took on a grimmer character. But his unhappiness gave him a new subject for his verse. The elegant couplets in which he had penned his love poetry became in his poems from exile the medium for lament.

We can now see that Ovid's love poetry is too light-hearted to deserve serious disapproval, and it may seem strange that Ovid, who unlike any of the other major Augustan poets was a respectable married man, should have become the victim of the Emperor's anger. But in seeming to encourage adultery, which Augustus had made a criminal offence, and in sending up traditional Roman stories (like that of Romulus and the Sabine women), Ovid was asking for trouble. Even now his love poetry has a *risqué* element, a whiff of sexual excitement, as the poet pursues love to the exclusion of everything else.

The great passion of Ovid's life was in fact poetry. We hope that his poignant tribute to his fellow love-poet Tibullus will prove an appropriate conclusion to a book dedicated to the Golden Age of Latin literature.

CAESAR

Outline of Caesar's life

BC

100	Caesar born
87	His father dies
81–78	He serves in the army in Asia
78	Sulla dies; Caesar returns to Rome
77	He prosecutes Dolabella
76	He is captured by pirates, en route to Rhodes
68	Caesar quaestor, serves in Further Spain
65	Caesar aedile
62	Caesar praetor
61	He governs Further Spain
60	He stands for the consulship. Opposed by the nobles, he forms the First Triumvirate with Pompey and Crassus
59	Caesar consul; he forces his measures through by violence
58–50	Caesar governor of Cisalpine and Transalpine Gaul The conquest of Gaul
55	First expedition to Britain (late August)
54	Second expedition to Britain (early July)
53	Crassus defeated and killed at Carrhae (Parthia)
51	Pompey joins the nobles against Caesar
49	Caesar invades Italy (crosses Rubicon 10 January). Civil War: Caesar against Pompey and the nobles
48	Caesar appointed dictator. He defeats Pompey at Pharsalus (in Greece) and follows him to Egypt
48–7	Caesar in Alexandria (affair with Cleopatra)
46	Battle of Thapsus; Caesar defeats Republican forces in Africa. He is elected dictator for ten years
45	Battle of Munda; Caesar defeats the sons of Pompey in Spain
44	Caesar is elected dictator for life. Ides of March: he is assassinated – conspiracy led by Brutus and Cassius

CHAPTER I

The young Caesar

Gāius Iūlius Caesar, nōbilissimō nātus genere, orīginem ab Iūlō, Aenēae fīliō, dēdūxit. cum mortua esset amita eius Iūlia, in laudātiōne fūnebrī dē eius et patris suī orīgine sīc dīxit:

'amitae meae Iūliae māternum genus ab rēgibus ortum,
5 paternum cum dīs immortālibus coniūnctum est. nam ab Ancō Martiō sunt Marciī rēgēs, quō nōmine fuit māter; ā Venere Iūliī, cuius gentis familia est nostra.' avunculus eius cōnsulātum adeptus est, pater praetūram; sed cum Caesar quīndecim annōs nātus esset, pater occidit, morbō correptus. māter igitur Aurēlia,
10 mulier ingeniō ēgregiō, eum ēdūcāvit summāque dīligentiā cūrāvit.

temporibus turbulentīs adolēvit. nam Sulla dictātor factus reīpūblicae dominābātur, quī Caesaris familiae inimīcus erat. Sulla enim senātūs auctōritātem restituerat, Caesaris autem
15 genus partibus favēbat populāribus. adulēscēns igitur Rōmā discessit ut Sullae inimīcitiam vītāret, mīlitābatque in Asiā, nōn sine glōriā; nam corōnā cīvicā dōnātus est, quod in expugnātiōne oppidī cuiusdam cīvem Rōmānum servāverat.

statim Sullā mortuō Rōmam reversus est et mox in rē
20 pūblicā versārī coepit. nam Dolabellam, virum īnsignem, quī cōnsul fuerat et prōcōnsul Asiae, repetundārum postulāvit. Dolabellā tamen absolūtō Rhodum sēcēdere cōnstituit ut apud Molōnem studēret, quī clārissimus tunc erat dīcendī magister.

hūc dum hībernīs iam mēnsibus nāvigat, prope
25 Pharmacussam, īnsulam Mīlētō vīcīnam, ā praedōnibus captus est mānsitque apud eōs quadrāgintā diēs cum ūnō medicō et servīs duōbus; nam comitēs statim dīmīserat ad argentum expediendum quō redimerētur. quī cum rediissent, argentō plūrimō praedōnibus numerātō, līberātus in lītore Asiae
30 expositus est. sine morā Mīlētum festīnāvit nāvibusque comparātīs ut praedōnēs persequerētur profectus est. assecūtus est eōs prope Pharmacussam adhūc cessantēs māiōremque partem eōrum cēpit. captīvōs eō suppliciō affēcit quod saepe illīs per iocum minātus erat; omnēs enim crucī suffīxit.

35 paucīs post annīs cursum honōrum iniit et quaestor factus missus est in Hispāniam Ulteriōrem. ōlim cum Gādēs advēnisset iūs dīcendī causā, Alexandrī Magnī imāgine cōnspectā ingemuit, commōtus quod ipse nōndum quicquam memorābile ēgisset eā aetāte quā Alexander orbem terrārum iam subēgisset. ubi

Sulla

Alexander the Great on horseback

2	**amita eius** his aunt. Julia was the widow of Marius.
3	**laudātiōne fūnebrī** funeral laudation, i.e. the speech he made in praise of her at her funeral.
5–6	Ancus Marcius was the fourth king of Rome; Julia's mother's family were called Marcii Reges.
6	**ā Venere:** according to tradition, Venus was the mother of Aeneas.
7–8	**cōnsulātum adeptus est** won the consulship.
10	**ingeniō ēgregiō** of remarkable character.
12	**turbulentīs** stormy. **Sulla dictātor:** over the last fifty years the authority of the Senate had been repeatedly challenged by popular leaders. This struggle for power culminated in civil war, when Sulla marched on Rome and drove out Marius, the great general and leader of the popular party (88 B.C.). While Sulla was fighting wars in the East, the **populārēs** again seized power. When Sulla returned to Italy with his army, he defeated the **populārēs** in a bitter war. He was then made dictator (81 B.C.) and restored the authority of the Senate by legislation.
17	**corōnā cīvicā dōnātus est** he was awarded the civic crown. This was a wreath of oak leaves, awarded to those who saved the life of a fellow citizen in battle.
19–20	**in rē pūblicā versārī** to take part in public life.
21	**repetundārum postulāvit** he prosecuted for extortion. Extortion means maladministration of a province. It was not unusual for an ambitious young man to make his mark by prosecuting one much his senior. Caesar was a gifted speaker.
22	**absolūtō** having been acquitted. **sēcēdere** to retire.
23	**dīcendī magister** teacher of speaking.
24	**hībernīs mēnsibus** in the winter months.
25	**ā praedōnibus** by pirates. These were a considerable menace at this time in the Eastern Mediterranean.
27–8	**ad argentum expediendum** to get the money.
28	**quō redimerētur** with which he might be ransomed = for his ransom.
29	**numerātō** having been paid.
32	**cessantēs** lingering.
33–4	**eō suppliciō** with that punishment. **per iocum** jokingly. While Caesar had been in their hands, he had treated the pirates with cheerful contempt.
34	**crucī suffīxit** fixed to a cross = crucified. Crucifixion was the usual way of executing violent criminals.
35	**cursum honōrum** the career of offices. Quaestorship was the lowest rung on the ladder.
36	**Gādēs** Cadiz.
37	**iūs dīcendī causā** to administer justice. **ingemuit** he groaned.
38–9	**eā aetāte:** Alexander the Great had conquered the known world by the age of thirty-one; Caesar was thirty-four when he was quaestor.

40 prīmum licuit, Rōmam rediit ut māiōra susciperet.

aedīlis creātus lūdōs magnificōs ēdidit quibus favōrem populī sibi conciliāret. in hīs annīs identidem dēclārābat sē nōbilibus adversārī, populāribus autem studēre, ita ut nōbilium et invidiam et suspiciōnem in sē movēret.

45 praetūrā cōnfectā in Hispāniam rūrsus missus est ut prōvinciam administrāret. illīc exercituī prīmum praefectus ducem sē praebuit ēgregium; gentibus enim Hispānīs usque ad mare Atlanticum subāctīs imperātor ā cōpiīs appellātus est triumphumque meruit. pācātā prōvinciā Rōmam rediit ad

50 cōnsulātum petendum.

nōbilēs tamen veritī nē Caesar auctōritātem suam īnfringeret, omnibus modīs eī adversābantur. ille igitur societātem clam iniit cum Pompēiō Crassōque ut inter sē cōnsociātī nōbilium potentiae resisterent omniaque ad suam

55 voluntātem regerent.

Kalendīs Iānuāriīs Caesar cōnsulātum iniit. prīmam rogātiōnem ad senātum rettulit. cum tamen nōbilēs eī obstārent, ad populum rem prōtinus tulit. cum collēga eius ōrātiōnem habēre cōnārētur quā rogātiōnem dissuādēret, dē rōstrīs ā turbā

60 dēiectus domum fugere coāctus est. ex eō tempore domī sē clausit, veritus nē ab operīs Caesaris occīderētur. Caesar omnia in rē pūblicā ūnus administrāvit; lēgēs quāscumque volēbat per plēbem tulit; sī quis eī resistēbat, vī et minīs dēterrēbātur. ūnus sōlus Caesarī palam adversārī audēbat, iuvenis quīdam Cūriō

65 nōmine.

Cicerō dē rēpūblicā dēspērāvit. mēnse Quīntīlī sīc ad Atticum scrībit:

scītō nihil umquam fuisse tam īnfāme, tam turpe, tam omnibus generibus, ordinibus, aetātibus offēnsum quam hunc

70 statum quī nunc est. illī nēminem tenent voluntāte; vereor nē metū necesse sit eīs ūtī . . . populī sēnsūs maximē theātrō et spectāculīs perspectus est . . . lūdīs Apollināribus Dīphilus tragoedus in Pompēium palam invectus est;
'nostrā miseriā tū es Magnus – '

75 mīliēns coāctus est dīcere;
'eandem virtūtem istam veniet tempus cum graviter gemēs' tōtius theātrī clāmōre dīxit. Caesar cum vēnisset mortuō plausū, Cūriō fīlius est īnsecūtus. huic maximus datus est plausus. tulit haec Caesar graviter.

80 haec scrīpsī properāns et timidē. posthāc ad tē aut, sī fidēlem habeō cui epistulam dem, scrībam plānē omnia, aut, sī obscūrē scrībam, tū tamen intellegēs.

sīc Caesar reīpūblicae tōtum annum dominātus omnia ad voluntātem suam administrābat. in prīmīs cum senātus eum

A tragoedus

39–40	**ubi prīmum licuit** as soon as he was allowed.
41	**aedīlis** this was the second rung on the ladder. The post was largely administrative and involved responsibility for the entertainments laid on at the public games.
42	**identidem** again and again.
44	**in sē movēret** roused against himself.
45	**praetūrā:** the praetorship was a senior magistracy. Praetors could command armies and were attended by six lictors. In Rome they presided over the law courts. After their year of office in Rome they were usually sent out to govern provinces as propraetors.
47	**usque ad** right up to.
48	**imperātor:** a general was given this title only after he had been hailed imperātor by his victorious troops. The Senate then decided whether the victory merited a triumph in Rome.
52	**īnfringeret** he might damage, undermine.
53	**societātem** alliance. **cum Pompēiō Crassōque:** Pompey was at this time the most powerful man in Rome (he had won great victories in the East); Crassus was the richest.
53–4	**inter sē cōnsociātī** united with each other.
54–5	**ad suam voluntātem** to their own wishes, as they liked.
56	**Kalendīs Iānuāriīs** on 1st January.
57	**rogātiōnem** bill. This was a proposal to be put before an assembly of the people. By tradition, bills were submitted to the Senate for discussion before they were put to the people.
58	**collēga** his colleague, i.e. the other consul.
61	**ab operīs** by the gangs.
63	**sī quis** if anyone.
66	**mēnse Quīntīlī** in the month of July.
68	**scītō** know, let me tell you; the imperative of **sciō.** **īnfāme** disgraceful.
69	**ordinibus** classes (of people). **offēnsum** offensive to, unpopular with.
70	**statum** state of affairs. **illī:** i.e. the triumvirs.
70–1	**nē metū . . . ūtī** = **nē eīs necesse sit metū ūtī.**
72	**lūdīs Apollināribus:** the games of Apollo took place between 6 and 13 July each year and would have included plays and spectacles such as gladiatorial shows.
72–3	**Dīphilus tragoedus** Diphilus the tragic actor. **in Pompēium palam invectus est** openly attacked Pompey. Pompey, since his victories in the East, was known as Pompeius Magnus or simply Magnus.
76	**eandem . . . gemēs** = **veniet tempus cum eandem virtūtem istam graviter gemēs:** we do not know what play these lines come from. **gemō, gemere** lament.
78	**Cūriō fīlius:** i.e. the young Curio.
78–9	**tulit graviter** took badly, was angry at.
80–2	Cicero was evidently afraid that under the tyranny of the Triumvirs his letters might be opened and read by spies. **cui dem** to whom I may give . . . .
84	**in prīmīs** in particular. **senātus:** the Senate normally decided which provinces should be allotted to the consuls and praetors. In allotting Caesar an unimportant province which did not require an army, they hoped to clip his wings. Caesar's answer was to get a tribune to propose in the Concilium Plebis that these arrangements should be changed.

85　　prōvinciae parvae praefēcisset imbellīque, populus ē plēbiscītō
　　　Galliam Cisalpīnam cum tribus legiōnibus eī assignāvit; deinde
　　　Galliam Trānsalpīnam imperiō eius addidit ipse senātus
　　　Pompēiō auctōre. cōnsulātū cōnfectō Caesar ad prōvinciam
　　　profectus est quam novem per annōs obtinēbat; quō in tempore
90　　omnem Galliam quae montibus Pȳrēnaeīs Alpibusque,
　　　flūminibus Rhēnō ac Rhodanō continētur in formam prōvinciae
　　　redēgit; Germānōs quī trāns Rhēnum incolunt aggressus
　　　maximīs affēcit clādibus; aggressus est et Britannōs, ignōtōs
　　　anteā, superātīsque pecūniās et obsidēs imperāvit. per tot
95　　successūs clādem vix ūllam sustinuit. quibus rēbus gestīs tantum
　　　glōriae tantāsque opēs sibi peperit, ut nōn modo nōbilēs sed ipse
　　　Pompēius metuerent nē Rōmam regressus reīpūblicae sōlus
　　　dominārētur.

Caesar's campaigns in Gaul

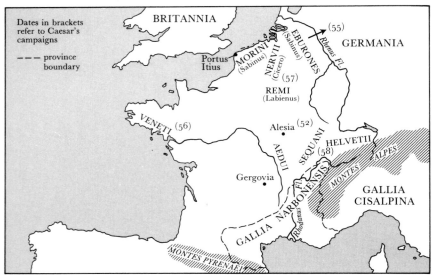

A barbarian warrior lies dead

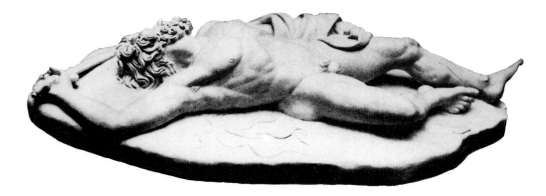

Decisions of the plebs were called **'plēbiscīta'**; these could override decisions of the Senate (**senātūs cōnsulta**).

86   **assignāvit** assigned.

88   **Pompēiō auctōre** on Pompey's proposal.

90   **omnem Galliam:** see map.

92   **redēgit** he reduced, brought.

93   **et Britannōs** the Britons also.

94   **superātīs** when they were conquered.   **pecūniās** money = taxes.

96   **peperit** won.

96–7   **ipse Pompēius:** Pompey and Caesar gradually drifted apart; Pompey had married Caesar's daughter Julia, but she had died in 54 B.C. Crassus had been killed at Carrhae in 53. Pompey became more and more nervous of Caesar's intentions and the nobles took advantage of this to draw him onto their side.

NB Principal parts of verbs are from now on given as follows: **amō, -āre, amāvī, amātum**. The fourth part (**amātum**) is called the supine and its use will be explained later. The supine is given because intransitive verbs have supines and from this part you can form the future participle, e.g. **veniō, -īre, vēnī, ventum**; future participle **ventūrus**.

administrō (1)	I manage, govern
aggredior, aggredī, aggressus	I attack
causa, -ae, *f.*	cause, reason; law case
causā	by reason of, for the sake of
dominor (1) + dative	I dominate, rule
faveō, -ēre, fāvī, fautum + dative	I favour, support
favor, favōris, *m.*	favour, support
imāgō, imāginis, *f.*	image, picture, statue
medicus, -ī, *m.*	doctor
mercātor, mercātōris, *m.*	merchant
nātiō, nātiōnis, *f.*	tribe
obses, obsidis, *c.*	hostage
opēs, opum, *f.pl.*	wealth, resources
quam celerrimē	as quickly as possible
quam prīmum	as soon as possible
reperiō, -īre, repperī, repertum	I find out, I find
sentiō, -īre, sēnsī, sēnsum	I feel, I perceive
sēnsus, -ūs, *m.*	feeling
sententia, -ae, *f.*	opinion, decision, vote
voluntās, voluntātis, *f.*	will, good will

The **cursus honōrum**

quaestor	**quaestūra**	quaestorship
aedīlis	**aedīlitās**	aedileship
praetor	**praetūra**	praetorship
cōnsul	**cōnsulātus**	consulship

G Word building

Make sure that you know the following compounds of **dō**, **dare**, **dedī**, **datum**; note that **dō** is first conjugation but all its compounds are third.

abdō, abdere, abdidī, abditum	I put away, hide	**ēdō** etc.	I give out
addō etc.	I put to, add	**perdō** etc.	I lose, waste, destroy
condō etc.	I put together, found	**prōdō** etc.	I betray
dēdō etc.	I give up, surrender	**reddō** etc.	I give back, return
		trādō etc.	I hand over, hand down

Give English derivatives from as many of these compounds as you can.

G The gerund

Molōn clārissimus tunc erat <u>dīcendī</u> magister
Molon was then the most famous master of speaking.
<u>dīcendī</u> is the genitive of a verbal noun which grammarians call the
gerund.
The gerunds of the four conjugations are:

1 **amandum** loving *2* **monendum** warning *3* **regendum** ruling
4 **audiendum** hearing

They are not used in the nominative; the other case endings are the
same as those of the singular of **bellum**:

accusative	**amandum**	loving	*dative*	**amandō**	to/for loving
genitive	**amandī**	of loving	*ablative*	**amandō**	by loving

The accusative of the gerund is often used after **ad** to express purpose:
Rōmam rediit ad cognōscendum quid accidisset
He returned to Rome to find out what had happened.
In the genitive it is dependent on another noun:
magister dīcendī master of speaking; **ars legendī** the art of reading;
signum prōgrediendī the signal of advancing (but we say 'for
advancing'); **effugiendī occāsiō** an opportunity for escaping.
The genitive is also used with **causā** = for the sake of, to express
purpose:
Rhodum nāvigāvit Molōnem audiendī causā
He sailed to Rhodes for the sake of hearing (= to hear) Molon.
It occurs occasionally in the dative, e.g.
hic locus idōneus est dormiendō
This is a suitable place for sleeping.
It occurs quite commonly in the ablative:
puer festīnandō domum ante noctem pervēnit
By hurrying the boy got home before night.
in studendō valdē dīligentem sē praebuit
He showed himself very industrious in studying.

Exercise 1.1

Translate

1 Caesar Rhodum nāvigāvit ad studendum apud Molōnem.
2 ā praedōnibus captus comitēs Mīlētum mīsit argentum comparandī causā.
3 līberātus statim nāvēs parāvit praedōnēs persequendī causā.
4 celerrimē nāvigandō praedōnēs assecūtus est.
5 Caesar societātem cum Pompēiō Crassōque iniit nōbilibus resistendī causā.
6 in Caesarī adversandō iuvenis sē fortissimum praebuit.
7 dux mīlitibus signum prōgrediendī dedit.
8 senex nūllam occāsiōnem āmīsit amīcōs adiuvandī.
9 iuvenis cōnsilium cēpit domum statim redeundī*.
10 statim proficīscendō domum eōdem diē advēnit.

*NB The gerund of **eō** and compounds is **eundum**, **redeundum** etc.

Exercise 1.2

Translate into Latin

1 I learnt the art of speaking in Rome.
2 Then I went to Athens to study in the Academia.
3 By working hard I learnt many things there.
4 At last I returned to Rome to see my parents.
5 When I was in Rome, I had the opportunity of hearing the best speakers.

The following passage and subsequent chapters are based on or quoted from Caesar's Commentāriī dē Bellō Gallicō. This is a kind of war diary, in which he describes his campaigns in seven books, one for each year from 58 to 52 BC. He writes in the third person, calling himself Caesar, and appears to be giving a straightforward account of a patriotic Roman fighting necessary wars. The books actually contain an element of propaganda. Caesar writes for his public in Rome to justify the continual extension of his conquests, and according to his account he never makes a mistake. In fact, as we shall see from his account of his invasions of Britain, he did make some mistakes. But it is also clear that he was a brilliant general and he writes extraordinarily clear and vivid accounts of his campaigns.

By the late summer of 55 BC Caesar had overrun all Gaul from the Alps to the Pyrenees: He had also, this year, made a raid into Germany over the Rhine. Although it was late in the campaigning season (he actually sailed on 26 August), he decided to make an expedition to Britain, as the Britons had continually sent help to the Gauls.

Exercise 1.3

Translate the first paragraph and answer the questions on the second

iam exiguā parte aestātis reliquā, tamen Caesar in Britanniam
proficīscī contendit, quod omnibus ferē Gallicīs bellīs
intellegēbat inde sumministrāta esse auxilia, et sī tempus annī ad
bellum gerendum dēficeret, tamen magnō sibi ūsuī fore
5   arbitrābātur, sī modo īnsulam adiisset et genus hominum
perspexisset, loca, portūs, aditūs cognōvisset; quae ferē omnia
Gallīs erant incognita. itaque vocātīs ad sē undique
mercātōribus, neque quae aut quantae nātiōnes incolerent,
neque quem ūsum bellī habērent neque quī essent ad māiōrum
10 nāvium multitūdinem idōneī portūs reperīre poterat.
    ad haec cognōscenda C. Volusēnum cum nāve longā

Two long ships

praemittit. huic mandat ut explōrātīs omnibus rēbus ad sē quam
prīmum revertātur. ipse cum omnibus cōpiīs in Morinōs
proficīscitur, quod inde erat brevissimus in Britanniam trāiectus.
15 interim cōnsiliō eius cognitō et per mercātōrēs perlātō ad
Britannōs, ā complūribus cīvitātibus ad eum lēgātī veniunt quī
pollicentur obsidēs dare atque imperiō populī Rōmānō
obtemperāre. quibus audītīs, līberāliter pollicitus hortātusque ut
in eā sententiā permanērent, eōs domum remittit.

1 What instructions did Caesar give Volusenus?
2 Why did Caesar set out to the Morini?
3 How did the Britons find out his plans?
4 What did the British ambassadors say?
5 How did Caesar reply?
6 From this paragraph give an example of: an ablative absolute; an
   indirect command; a connecting relative.
7 In the last sentence (**quibus. . . .remittit**) seven out of the thirteen
   words have English derivatives; write down as many of these as you
   can and say what each English word means.
8 What were Caesar's motives for invading this year? What did he
   hope to achieve? Was Caesar wise to make an expedition at this
   time of year?

1	**exiguā** little.
3	**sumministrāta esse** had been provided.
4	**magnō sibi ūsuī fore** it would be of great use (= very useful) to him.
7	**incognita** unknown.
8–10	**neque quae . . . portūs:** these clauses all depend on **reperīre poterat.**
11	**nāve longā:** a long ship is a warship. **ad haec cognōscenda** to find out these things.
12	**huic mandat** he instructs him.
13	**in Morinōs:** see map on p. 14.
14	**trāiectus** crossing.
16	**complūribus** several. **cīvitātibus** states. Britain at this time was divided between a large number of independent tribes. **lēgātī** ambassadors.
18	**obtemperāre** (+ dative) to submit to.

Exercise 1.4

Translate into Latin

Although summer was nearly over, Caesar decided to set out for Britain at once. But because he could not discover from merchants which ports were suitable for large ships, he sent Volusenus to find out everything. He ordered him to return as quickly as possible. He himself, after collecting eighty ships, prepared to transport two legions to Britain.

'nostrā miseriā tū es Magnus'

CHAPTER II

The first invasion of Britain (late August 55 BC)

nāvibus circiter LXXX onerāriīs coāctīs, quod satis esse ad duās
trānsportandās legiōnēs exīstimābat, quod praetereā nāvium
longārum habēbat quaestōrī, lēgātīs praefectīsque distribuit.

5   hīs cōnstitūtīs rēbus, nactus idōneam ad nāvigandum
tempestātem, tertiā vigiliā nāvēs solvit. hōrā circiter diēī quārtā
cum prīmīs nāvibus Britanniam attigit atque ibi in omnibus
collibus expositās hostium cōpiās armātās cōnspexit. cuius locī
haec erat nātūra atque ita montibus angustīs mare continēbātur,
ut ex locīs superiōribus in lītus tēlum adigī posset. hūc ad
10 ēgrediendum nōn idōneum locum arbitrātus, dum reliquae nāvēs
convenīrent ad nōnam hōram in ancorīs exspectāvit. tum ventum
et aestum ūnō tempore nactus secundum, datō signō et sublātīs
ancorīs, circiter mīlia passuum septem ab eō locō prōgressus
apertō ac plānō lītore nāvēs cōnstituit.

15   at barbarī, cōnsiliō Rōmānōrum cognitō, praemissō
equitātū et essedāriīs, reliquīs cōpiīs subsecūtī, nostrōs nāvibus

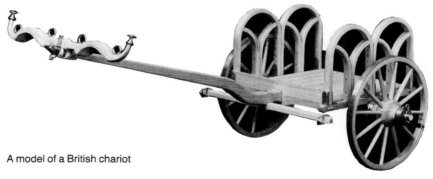

A model of a British chariot

ēgredī prohibēbant. erat ob hās causās summa difficultās, quod
nāvēs propter magnitūdinem nisi in altō cōnstituī nōn poterant.
mīlitibus autem magnō et gravī onere armōrum oppressīs simul
20 et dē nāvibus dēsiliendum erat et in flūctibus cōnsistendum et
cum hostibus pugnandum. illī tamen aut ex āridō aut paulum in
aquam prōgressī audācter tēla coniciēbant et equōs in nostrōs
incitābant.

    quod ubi Caesar animadvertit, nāvēs longās paulum
25 removērī ab onerāriīs nāvibus iussit et ad latus apertum hostium
cōnstituī atque inde fundīs, sagittīs, tormentīs hostēs summovērī;
quae rēs magnō ūsuī nostrīs fuit. nam barbarī inūsitātō genere
tormentōrum permōtī cōnstitērunt ac paulum pedem
rettulērunt. atque nostrīs mīlitibus cunctantibus, maximē

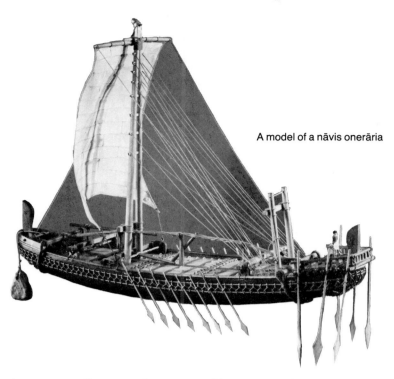

A model of a nāvis onerāria

1	**nāvibus onerāriīs** transport ships, troop carriers.
2–3	**quod praetereā . . . habēbat** what of warships he had besides = the warships which he had besides (the troop carriers). **quaestōrī** Caesar would have had with him one quaestor, two legates (= legionary commanders) and two prefects (= camp commandants).
4	**nactus** having obtained.
5	**tertiā vigiliā** the third watch (of the night). This would have been from midnight to 2.30 a.m. He sailed from Boulogne and arrived at Britain at the fourth hour (about 9 a.m.) opposite the cliffs of Dover. At the ninth hour (about 3 p.m.) he sailed Eastwards about seven miles to the level beach between Deal and Walmer.
8–9	**haec . . . ut** such . . . that.
9	**tēlum adigī posset** missile(s) could be fired. **tēlum** is used of any weapon used for fighting at a distance.
10–11	**dum . . . convenīrent** until they should gather. The subjunctive expresses purpose.
16	**essedāriīs** charioteers. The Britons were famous for their war chariots.
17	**ob hās causās** for the following reasons.
19–21	**mīlitibus . . . dēsiliendum erat . . . et cōnsistendum . . . et pugnandum** the soldiers had to jump down and had to take up position and had to fight.
21	**ex āridō** from dry land.
26	**fundīs, sagittīs, tormentīs** slings, arrows, artillery. The sling was an effective weapon which could fire stones or lead bullets up to a range of about 100 yards. Roman artillery consisted of massive catapults; the largest could shoot a stone weighing 162 lb. up to 500 yards, but those mounted on ships were smaller.
27–8	**inūsitātō genere tormentōrum** the unfamiliar type of artillery = artillery, a type of weapon unfamiliar to them.

30 propter altitūdinem maris, is quī decimae legiōnis aquilam
    ferēbat, precātus deōs ut ea rēs legiōnī fēlīciter ēvenīret,
    'dēsilite,' inquit, 'mīlitēs, nisi vultis aquilam hostibus prōdere.
    ego certē meum reīpūblicae atque imperātōrī officium
    praestiterō.' hoc cum magnā vōce dīxisset, sē ex nāve prōiēcit
35 atque in hostēs aquilam ferre coepit. tum nostrī cohortātī inter sē
    nē tantum dēdecus admitterētur, ūniversī ex nāve dēsiluērunt.

    pugnātum est ab utrīsque ācriter. nostrī tamen quod neque
    ordinēs servāre neque firmiter īnsistere neque signa subsequī
    poterant, magnopere perturbābantur. hostēs autem, nōtīs
40 omnibus vadīs, ubi aliquōs singulārēs ex nāve ēgredientēs
    cōnspexerant, incitātīs equīs eōs impedītōs adoriēbantur.

    quod cum animadvertisset Caesar, scaphās longārum
    nāvium mīlitibus complērī iussit et, quōs labōrantēs cōnspexerat,
    hīs subsidia mittēbat. nostrī, simul atque in āridō cōnstitērunt,
45 suīs omnibus cōnsecūtīs, in hostēs impetum fēcērunt atque eōs in
    fugam dedērunt.

    hostēs proeliō superātī, simul atque sē ē fugā recēpērunt,
    statim ad Caesarem lēgātōs dē pāce mīsērunt; obsidēs datūrōs
    esse quaeque imperāvisset sē factūrōs pollicitī sunt. Caesar
50 questus quod, cum ultrō in continentem lēgātīs missīs pācem ā sē
    petiissent, bellum sine causā intulissent, sē ignōscere
    imprūdentiae dīxit obsidēsque imperāvit; quōrum illī partem
    statim dedērunt, partem ex longinquiōribus locīs arcessītam
    paucīs diēbus sē datūrōs esse dīxērunt.

55 eādem nocte accidit ut esset lūna plēna, quī diēs maritīmōs
    aestūs maximōs in Oceanō efficere cōnsuēvit, nostrīsque id erat
    incognitum. ita ūnō tempore et longās nāvēs, quās Caesar in
    āridum subdūxerat, et onerāriās, quae ad ancorās erant
    dēligātae, tempestās adflīctābat. complūribus nāvibus frāctīs,
60 cum reliquae essent ad nāvigandum inūtilēs, magna tōtīus
    exercitūs perturbātiō facta est. neque enim nāvēs aliae erant
    quibus reportārī possent et omnia dēerant quae ad reficiendās
    nāvēs erant ūsuī et frūmentum in hīs locīs in hiemem prōvīsum
    nōn erat.

65 quibus rēbus cognitīs prīncipēs Britannōrum, quī post
    proelium ad Caesarem convēnerant, inter sē collocūtī, cum et
    nāvēs et frūmentum Rōmānīs dēesse intellegerent et paucitātem
    mīlitum ex castrōrum exiguitāte cognōscerent, rūrsus
    coniūrātiōne factā, paulātim ex castrīs discēdere et suōs clam ex
70 agrīs dēdūcere coepērunt.

    at Caesar, etsī nōndum eōrum cōnsilia cognōverat, ad
    omnēs cāsūs subsidia comparābat. nam et frūmentum ex agrīs
    cotīdiē in castra cōnferēbat et eārum nāvium quae gravissimē
    afflīctae erant māteriā et aere ad reliquās reficiendās ūtēbātur.
75 itaque cum summō studiō ā mīlitibus rēs administrārētur, effēcit

A scapha

33	**certē** at least.
34	**praestiterō** will have done.
35	**cohortātī inter sē** encouraging each other.
36	**tantum dēdecus**: the disgrace would be the loss of the standard, if they did not follow the standard-bearer.
37	**pugnātum est** it was fought = the battle was fought.
38	**firmiter īnsistere** stand firmly, get a firm foothold. **signa** each unit down to the century and maniple (subdivision of the century) had its own standard. In battle you followed your standard-bearer, who could signal whether to advance, retreat or rally. If the standard-bearer was invisible or not in position, chaos could result.
40	**vadīs** shallows. **singulārēs** on their own.
42	**scaphās** ship's boats.
43	**labōrantēs** labouring = in difficulties.
44	**subsidia** help.
50	**ultrō** of their own accord.
53	**longinquiōribus** more distant.
55–6	**maritīmōs aestūs** the sea tides. **Oceanō** the Ocean (here the Atlantic) is the sea outside the Mediterranean. There are virtually no tides in the Mediterranean, but it is impossible to believe that the Romans did not know about the Atlantic tides, especially as Caesar had fought a naval war against the Veneti in Brittany the year before. **cōnsuēvit** is accustomed to.
62–3	**quae erant ūsuī** which were useful for.
63	**in hiemem** for winter.
68	**ex exiguitāte** from the small size.
69	**coniūrātiōne** conspiracy, plot.
72	**subsidia** protection (*n. pl*).
74	**māteriā** timber. **aere** bronze.
75	**cum** since.
75–6	**effēcit ut** he brought it about that . . . , he achieved that . . . .

ut XII nāvibus āmissīs reliquīs nāvigārī commodē posset.

dum ea geruntur, ūnā legiōne ex cōnsuētūdine missā ut frūmentum colligeret, eī quī prō portīs castrōrum in statiōne erant Caesarī nūntiāvērunt pulverem plūrimum in eā parte vidērī
80 quam in partem legiō iter fēcisset. Caesar cohortēs quae in statiōnibus erant sēcum in eam partem proficīscī, reliquās armārī et cōnfestim sē subsequī iussit. cum paulō longius ā castrīs prōcessisset, suōs ab hostibus premī atque aegrē sustinēre animadvertit.

85 nam quod omnī ex reliquīs partibus dēmessō frūmentō pars ūna erat reliqua, suspicātī hostēs hūc nostrōs esse ventūrōs, noctū in silvīs dēlituerant. tum nostrōs dispersōs, dēpositīs armīs in metendō occupātōs, subitō adortī, paucīs interfectīs, reliquōs perturbāverant; simul equitātū atque essedīs circumdederant.

90 perturbātīs nostrīs novitāte pugnae Caesar opportūnissimō tempore auxilium tulit; namque eius adventū hostēs cōnstitērunt, nostrī sē ex timōre recēpērunt. quō factō Caesar arbitrātus tempus ad proelium committendum nōn idōneum esse, suō sē locō continuit et, brevī tempore intermissō, in castra
95 legiōnēs redūxit.

A notice at Deal records Caesar's landing

altus-a-um	high, deep	**incitō** (1)	I urge on
altum, -ī, *n.*	the deep, deep water	**interim**	meanwhile
altitūdo, altitūdinis, *f.*	height; depth	**paulātim**	little by little
aquila, -ae, *f.*	eagle; legionary standard	**perturbō** (1)	I throw into confusion
cōgō, -ere, coēgī, coāctum	I gather, collect; I compel		
cōnstituō, -ere, cōnstituī, cōnstitūtum	I set up, position; I decide	**perturbātiō, perturbātiōnis**, *f.*	confusion
cunctor (1)	I delay		
decus, decoris, *n.*	honour, glory	**pulvis, pulveris**, *m.*	dust
dēdecus, dēdecoris, *n.*	disgrace		
dēsiliō, -īre, dēsiluī	I jump down	**ūsus, ūsūs**, *m.*	use
etsī	although	**ūtilis-e**	useful
facultās, facultātis, *f.*	opportunity	**inūtilis-e**	useless
idōneus-a-um	suitable		

76	**nāvigārī commodē posset** it could be sailed properly with the rest = the rest were fit for sailing.
77	**ex cōnsuētūdine** according to custom.
78	**in statiōne** on guard duty.
82	**cōnfestim** at full speed.
83	**suōs** his men. **aegrē sustinēre** were scarcely withstanding (the attack).
85	**dēmessō** having been reaped.
87	**dēlituerant** had hidden.
88	**in metendō** in reaping. **adortī** having attacked.
90	**novitāte pugnae** by the novelty of the fighting = by the unfamiliar nature of the fighting. The Romans were not familiar with war chariots and the Britons' tactics.
94	**suō sē locō continuit** held himself in his own position = he stood fast. **brevī tempore intermissō** a short space of time having been allowed to pass = after a short interval.

Compounds of **sum**:

absum, abesse, āfuī	I am away from, distant from
adsum, adesse, adfuī	I am present, I am near
dēsum, dēesse, dēfuī	I am lacking; I fail
īnsum, inesse, īnfuī	I am in, I am among
praesum, praeesse, praefuī	I am in charge of
prōsum, prōdesse, prōfuī	I am useful to, I benefit
supersum, superesse, superfuī	I remain, survive

The gerundive

omnia dēsunt quae ūtilia sunt <u>ad nāvēs reficiendās</u>.
What is the meaning of the phrase underlined?

Caesar profectus est <u>ad Britanniam oppugnandam</u>.

In these phrases notice that **reficiendās** agrees with **nāvēs** and **oppugnandam** with **Britanniam**. **reficiendās** and **oppugnandam** are passive verbal adjectives, called *gerundives*. The first example means literally:
'Everything is lacking which is useful for the ships to-be-repaired.'

The gerundives of the four conjugations are:
1 **amandus-a-um** to be loved   *2* **monendus-a-um** to be warned
3 **regendus-a-um** to be ruled   *4* **audiendus-a-um** to be heard

Thus: **puer monendus est** The boy is to be warned.
      parentēs amandī sunt Parents are to be loved.
But **ad Britanniam oppugnandam** (= to Britain to be attacked) cannot be translated literally; it simply means 'to attack Britain'.

Exercise 2.1

Translate
1 Caesar Volusēnum praemīsit ad haec cognōscenda.
2 Caesar LXXX nāvēs coēgit ad Britanniam oppugnandam.
3 exīstimābat nāvēs LXXX satis esse ad duās trānsportandās legiōnēs.
4 Britannī in aquam prōgrediēbantur ad Rōmānōs repellendōs.
5 auxiliō mittendō Caesar suōs labōrantēs adiūvit.
6 Britannī victī lēgātōs mīsērunt ad pācem petendam.
7 frūmentō cotīdiē in castra cōnferendō satis habēbimus ad obsidiōnem sustinendam.
8 aquilā hostibus prōdendā magnum dēdecus admittētis.
9 impetū ācrī faciendō nostrī hostēs in fugam vertērunt.
10 castrīs optimē mūniendīs nāvēs cōnservābimus.

Exercise 2.2

In the following sentences use gerundives to translate the phrases underlined
1 The Britons have sent ambassadors to ask for peace.
2 By sending help to the Gauls they have shown themselves enemies of the Roman people.
3 We have enough ships (= enough of ships) to transport three legions.
4 We will sail at once to attack the Britons.
5 By doing this we shall make Gaul safer.

Exercise 2.3

Translate the first paragraph and answer the questions below
This passage follows straight on from the end of the preceding narrative.

interim barbarī nūntiōs in omnēs partēs dīmīsērunt
paucitātemque nostrōrum mīlitum suīs praedicāvērunt et
dēmōnstrāvērunt quanta facultās darētur praedae faciendae
atque in perpetuum sē līberandī, sī Rōmānōs castrīs expulissent.
5 hīs rēbus celeriter magnā multitūdine peditātūs equitātūsque
coāctā ad castra vēnērunt.
     Caesar legiōnēs in aciē prō castrīs cōnstituit. commissō
proeliō, diūtius nostrōrum mīlitum impetum hostēs ferre nōn
potuērunt ac terga vertērunt. quōs nostrī tantō spatiō secūtī
10 quantum cursū et vīribus efficere potuērunt, complūrēs ex eīs
occīdērunt, deinde omnibus longē lātēque aedificiīs incēnsīs, sē
in castra recēpērunt.

eōdem diē lēgātī ab hostibus missī ad Caesarem dē pāce
vēnērunt. hīs Caesar numerum obsidum quem anteā
15   imperāverat duplicāvit eōsque in continentem addūcī iussit. ipse
idōneam tempestātem nactus paulō post mediam noctem nāvēs
solvit, quae omnēs incolumēs ad continentem pervēnērunt.

2	**praedicāvērunt** announced.
9–10	**tantō spatiō quantum ... potuērunt** for so much space (= distance) as théy could achieve by running and strength = as far as their strength and speed allowed them.
11	**longē lātēque** far and wide.
15	**duplicāvit** doubled.
16	**nactus** having obtained.

1 How did the Romans follow up their victory?
2 What orders did Caesar give to the British ambassadors?
3 Give one example each from these paragraphs of: an ablative
   absolute; a connecting relative; a reflexive pronoun.
4 From what Latin adjective are the following formed: **paucitātem**
   (1.2); **multitūdine** (1.5)? Give an English word derived from each.
5 Summarize the story of this chapter as a whole in not more than 250
   words.
6 In this chapter:
   (a) In what respects does Caesar show himself a good general? Can
   you find any instances of incompetence?
   (b) How determined was the opposition of the Britons?

Exercise 2.4

Translate into Latin
When the Britons heard that Caesar had set out, they gathered their
forces and marched to the coast. And so when the Romans reached
the shore, they saw that the enemy were drawn up on all the hills.
Caesar ordered his men to sail to another place, where they could land
more easily. But the Britons, following them with all their forces,
fiercely attacked the Romans as they tried to land.

CHAPTER

The second invasion of Britain

*The first expedition to Britain began late in the campaigning season and
was little more than a reconnaissance. It was quickly over and Caesar
did not penetrate beyond the coastal area. He was hampered by lack of
cavalry (we have omitted the chapters in which he explains that the
cavalry transports were driven back to Gaul by bad weather and never
arrived in Britain at all). The following year he decided to make a large-
scale invasion. He instructed his legates to build a fleet of specially
designed transport ships. When he returned to Gaul from his annual
visit to Italy, he found that 600 transport ships had been built and
twenty-eight warships gathered. He ordered them all to assemble at
Boulogne and prepared to take five legions and 2,000 cavalry to
Britain. He sailed at the beginning of July, two months earlier than in
the previous year.*

    ad sōlis occāsum nāvēs solvit et lēnī Africō prōvectus mediā
circiter nocte ventō intermissō cursum nōn tenuit, et longius
dēlātus aestū, ortā lūce sub sinistrā Britanniam relictam
cōnspexit. tum rūrsus aestūs commūtātiōnem secūtus rēmīs
5   contendit ut eam partem īnsulae caperet quā optimum ēgressum
esse superiōre annō cognōverat. quā in rē admodum fuit mīlitum
virtūs laudanda, quī vectōriīs gravibusque nāvigiīs nōn
intermissō rēmigandī labōre longārum nāvium cursum
adaequāvērunt.
10   accessum est ad Britanniam omnibus nāvibus merīdiānō
ferē tempore neque in eō locō hostis vīsus est; sed, ut posteā
Caesar ex captīvīs cognōvit, cum magnae manūs eō
convēnissent, multitūdine nāvium perterritae ā lītore
discesserant ac sē in superiōra loca abdiderant.
15   Caesar expositō exercitū et locō castrīs idōneō captō, ubi ex
captīvīs cognōvit quō in locō hostium cōpiae consēdissent,
cohortibus decem ad mare relictīs, et equitibus trēcentīs, quī
praesidiō nāvibus essent, dē tertiā vigiliā ad hostēs contendit.
    noctū prōgressus mīlia passuum circiter XII hostium cōpiās
20   cōnspicātus est. illī equitātū atque essedāriīs ad flūmen prōgressī
ex locō superiōre nostrōs prohibēre et proelium committere
coepērunt. repulsī ab equitātū nostrō sē in silvās abdidērunt,
locum nactī ēgregiē et nātūrā et opere mūnītum. at mīlitēs

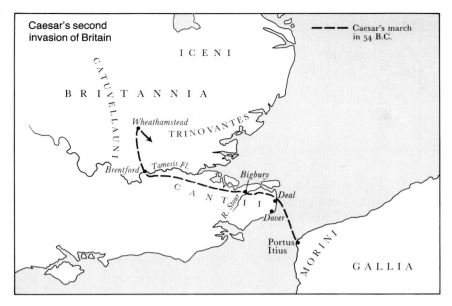

Caesar's second invasion of Britain

Caesar's march in 54 B.C.

ICENI

BRITANNIA

CATUVELLAUNI

Wheathamstead

TRINOVANTES

Brentford

Tamesis Fl.

Bigbury

Deal

CANTII

R. Stour

Dover

Portus Itius

MORINI

GALLIA

Line		
1	**ad sōlis occāsum** towards sunset. **lēnī Africō** by a gentle south-west wind.	
2	**ventō intermissō** when the wind dropped.	
3	**sub sinistrā** on the left = on the port beam.	
5	**ēgressum** from the noun **ēgressus, -ūs** getting out = landing place.	
6	**admodum** very much, extremely.	
7	**vectōriīs nāvigiīs** in transport ships.	
7–8	**nōn intermissō labōre** work not left off, i.e. they had no rest from the labour.	
9	**adaequāvērunt** equalled.	
10	**accessum est** it was approached = they approached.	
17–18	**quī praesidiō . . . essent** who might be for a guard to = to guard.	
20	**ad flūmen:** the river Stour, near Canterbury; see map. The speed of Caesar's operations was quite remarkable; since midday he had landed five legions (over 20,000 men) and 2,000 cavalry and set up camp. He had started in pursuit of the enemy at about 1 a.m.; he then marched twelve miles with his whole army except ten cohorts, engaged the enemy and took their fort.	

ad flūmen prōgressī ex locō superiōre (the river Stour, near Canterbury)

25   adiectō, locum cēpērunt eōsque ex silvīs expulērunt, paucīs
vulneribus acceptīs. sed eōs fugientēs longius Caesar prōsequī
vetuit, et quod locī nātūram ignōrābat et quod magnā parte diēī
cōnsūmptā mūnītiōnī castrōrum tempus relinquī volēbat.
        postrīdiē māne mīlitēs equitēsque mīsit ut eōs quī fūgerant
30   persequerentur. hīs aliquantum itineris prōgressīs, cum iam
extrēmī essent in cōnspectū, equitēs ā Q. Atriō ad Caesarem
vēnērunt quī nūntiārent superiōre nocte maximā coortā
tempestāte prope omnēs nāvēs adflīctās atque in lītore ēiectās
esse, quod neque ancorae fūnēsque subsisterent neque nautae
35   gubernātōrēsque vim tempestātis patī possent. hīs rēbus cognitīs
Caesar legiōnēs equitātumque revocārī iubet; ipse ad nāvēs
revertitur. cōnstituit omnēs nāvēs subdūcī et cum castrīs ūnā
mūnītiōne coniungī. in hīs rēbus circiter diēs X cōnsūmit.
subductīs nāvibus castrīsque ēgregiē mūnītīs eōdem unde
40   redierat proficīscitur.
        eō cum vēnisset, māiōrēs iam undique in eum locum cōpiae
Britannōrum convēnerant, summā imperiī bellīque
administrandī commūnī cōnsiliō permissā Cassivellaunō, cuius
fīnēs ā maritīmīs cīvitātibus flūmen dīvidit, quod appellātur
45   Tamesis, ā marī circiter mīlia passuum LXXX.
        Britanniae pars interior ab eīs incolitur quōs nātōs in īnsulā
ipsī memoriā prōditum dīcunt; maritīma pars ab eīs quī praedae
ac bellī īnferendī causā ex Belgiō trānsierant. ex eīs omnibus
longē sunt hūmānissimī quī Cantium incolunt, quae regiō est
50   maritīma omnis, neque multum ā Gallicā differunt
cōnsuētūdine. interiōrēs plērīque frūmenta nōn serunt sed lacte
et carne vīvunt pellibusque sunt vestītī. omnēs vērō sē Britannī
vitrō īnficiunt, quod caeruleum efficit colōrem, atque hōc
horridiōrēs sunt in pugnā aspectū; capillōque sunt prōmissō
55   atque omnī parte corporis rāsā praeter caput et labrum superius.
uxōrēs habent dēnī duodēnīque inter sē commūnēs, et maximē
frātrēs cum frātribus parentēsque cum līberīs.

omnēs nāvēs in lītore ēiectae sunt

24 **testūdine factā** forming a tortoise. This was a formation used in attacking a strongly fortified post; shields were locked over the heads of the soldiers (see picture).

24–5 **aggere adiectō** throwing a ramp up to . . .; **agger** means a pile of earth, stone etc. (it is used of the rampart of a camp); legionary soldiers carried entrenching tools for operations of this sort.

28 **mūnītiōnī castrōrum** a Roman army on the march through enemy territory built a full-scale fortified camp every day; this operation took several hours.

30 **aliquantum itineris** a little of the journey.

34 **fūnēs** cables. **subsisterent** held fast.

37 **omnēs nāvēs subdūcī** all the ships should be led down (to the shore) = beached. Caesar had been caught twice by tides and gales and now had no intention of allowing it to happen again.

42–3 **summā . . . administrandī** the whole of the command and managing the war (**summa** is a noun) = chief command and management of the war. The British tribes were usually at war with each other but in this crisis **commūnī cōnsiliō** they made Cassivellaunus commander-in-chief. He was a powerful and able man, king of the Catuvellauni, who lived in Hertfordshire, Middlesex and Oxfordshire.

46–7 **quōs . . . dīcunt** who they themselves say it is handed down by memory that they were born in the island = who according to native tradition were indigenous. In fact the Catuvellauni and other tribes as far north as the Parisii in East Yorkshire had crossed from the continent during the iron age and yet others came over both before and after Caesar's invasion.

49 **hūmānissimī** the most civilised. **Cantium** Kent.

50–1 **neque multum . . . cōnsuētūdine** and do not differ much from Gallic customs = and do not differ much from the Gauls in their customs.

51 **serunt** sow.

51–2 **lacte et carne** on milk and meat.

52 **pellibus** in skins (of animals).

53 **vitrō** with woad. This is a plant which produces a blue dye. **sē īnficiunt** dye themselves. **caeruleum** blue.

53–4 **hōc horridiōrēs . . . aspectū** by (doing) this they are more terrifying to see. **capillō sunt prōmissō** they are of long hair = they have long hair.

55 **rāsā** shaven. **praeter labrum superius** except their upper lip, i.e. they wore moustaches.

56 **uxōrēs habent . . . commūnēs** ten or twelve men at a time have wives shared in common.

equitēs hostium essedāriīque ācriter proeliō cum equitātū
nostrō in itinere cōnflīxērunt, tamen ut nostrī omnibus partibus
60   superiōrēs fuerint atque eōs in silvās compulerint. posterō diē
procul ā castrīs in collibus hostēs cōnstitērunt rārīque sē
ostendere et lēnius quam prīdiē nostrōs equitēs lacessere
coepērunt. sed merīdiē cum Caesar pābulandī causā trēs legiōnēs
atque omnem equitātum mīsisset, repente ex omnibus partibus
65   ad pābulātōrēs advolāvērunt. nostrī ācriter in eōs impetū factō
reppulērunt, magnō numerō eōrum interfectō, neque fīnem
sequendī fēcērunt quoad hostēs praecipitēs ēgērunt. ex hāc fugā
prōtinus auxilia quae undique convēnerant discessērunt, neque
post id tempus unquam summīs cōpiīs hostēs nōbīscum
70   contendērunt.
    Caesar, cognitō cōnsiliō eōrum, ad flūmen Tamesim in fīnēs
Cassivellaunī exercitum dūxit; quod flūmen ūnō omnīnō locō
pedibus, atque hōc aegrē, trānsīrī potest. eō cum vēnisset,

flūmen ūnō locō pedibus trānsīrī
potest (the Thames at Brentford)

animum advertit ad alteram flūminis rīpam magnās esse cōpiās
75   hostium īnstrūctās. rīpa autem erat acūtīs sudibus praefīxīsque
mūnīta, eiusdemque generis sub aquā dēfīxae sudēs flūmine
tegēbantur. eīs rēbus cognitīs ā captīvīs trānsfugīsque, Caesar
praemissō equitātū cōnfestim legiōnēs subsequī iussit. sed eā
celeritāte atque eō impetū mīlitēs iērunt, cum capite sōlō ex aquā
80   exstārent, ut hostēs impetum legiōnum sustinēre nōn possent
rīpāsque dīmitterent ac sē fugae mandārent.
    Cassivellaunus, omnī dēpositā spē contentiōnis, dīmissīs
ampliōribus cōpiīs, itinera nostra servābat paulumque ex viā
discēdēbat locīsque impedītīs ac silvestribus sēsē occultābat, ac
85   cum equitātus noster līberius praedandī causā sē in agrōs
ēiēcerat, omnibus viīs sēmitīsque essedāriōs ex silvīs ēmittēbat et
magnō cum perīculō nostrōrum equitum cum eīs cōnflīgēbat.
    ā perfugīs cognōscit Caesar nōn longē ab eō locō oppidum
Cassivellaunī abesse silvīs palūdibusque mūnītum, quō satis

59	**cōnflīxērunt** fought. **tamen ut** but with the result that ....
62	**lēnius** more gently = less violently. **lacessere** to provoke.
63	**pābulandī causā: pābulor** I forage, collect food.
67	**quoad** until. **praecipitēs** headlong.
68	**auxilia** helps = reinforcements.
69	**summīs cōpiīs** with full forces. From now on Cassivellaunus used guerilla tactics. It looks as if Caesar, in sending three legions out to forage, was tempting Cassivellaunus into a trap; his allies melted away after this battle.
71	**Tamesis** the Thames.
73	**atque hōc aegrē** and at this point with difficulty. The place where Caesar crossed the Thames is probably Brentford.

Coin of
Cassivellaunus

oppidum Cassivellaunī (the defensive ditch at Wheathamstead)

74	**animum advertit** he turned his mind = he noticed, saw.
75	**acūtīs sudibus praefīxīsque** with sharpened stakes fixed in front (of it).
77	**ā trānsfugīs** from deserters.
78–9	**eā celeritāte atque eō impetū** with such speed and such dash. Caesar often uses parts of **is, ea, id** to look forward to a consequence **ut.**
79	**cum** although.
81	**sē fugae mandārent** gave themselves up to flight.
82	**contentiōnis** of set battle.
84	**sēsē = sē.**
85	**līberius** too freely.
86	**sēmitīs** paths.
88–9	**oppidum Cassivellaunī:** this was probably Wheathamstead in Hertfordshire (not far from St. Albans).

90   magnus hominum pecorisque numerus convēnerit. eō
     proficīscitur cum legiōnibus. locum reperit ēgregiē nātūrā atque
     opere mūnītum; tamen hunc duābus ex partibus oppugnāre
     contendit. hostēs paulīsper morātī mīlitum nostrōrum impetum
     nōn tulērunt sēsēque aliā ex parte oppidī iēcērunt. magnus ibi
95   numerus pecoris repertus, multīque in fugā sunt comprehēnsī
     atque interfectī.
          dum haec in eīs locīs geruntur, Cassivellaunus nūntiōs ad
     Cantium mittit atque rēgibus imperat utī coāctīs omnibus cōpiīs
     castra nāvālia dē imprōvīsō adoriantur atque oppugnent. eī cum
100 ad castra vēnissent, nostrī ēruptiōne factā, multīs eōrum
     interfectīs, suōs incolumēs redūxērunt.
          Cassivellaunus, hōc proeliō nūntiātō, tot dētrīmentīs
     acceptīs, vāstātīs fīnibus, maximē etiam permōtus dēfectiōne
     cīvitātum, lēgātōs dē dēditiōne ad Caesarem mittit. Caesar
105 obsidēs imperat et quid vectīgālis in annōs singulōs populō
     Rōmānō Britannia penderet cōnstituit. obsidibus acceptīs
     exercitum redūcit ad mare, nāvēs invēnit refectās. eīs dēductīs,
     quod et captīvōrum magnum numerum habēbat et nōnnūllae
     tempestāte dēperierant nāvēs, duōbus commeātibus exercitum
110 reportāre cōnstituit. ac sīc accidit utī ex tantō numerō nāvium
     neque hōc neque superiōre annō ūlla omnīnō nāvis quae mīlitēs
     portāret, dēsīderārētur. ipse secundā initā cum solvisset vigiliā,
     prīmā lūce terram attigit omnēsque incolumēs nāvēs perdūxit.

Although Caesar speaks in the final paragraph as if Britain were now a
Roman province, this large-scale expedition achieved nothing
permanent. The Augustan poets continually predict that Augustus will
add Britain to the Roman empire, but in fact it was Claudius who
initiated the conquest of Britain, almost exactly 100 years after Caesar's
expeditions (AD 43).

V

amplus-a-um	ample, large, spacious	**noctū**	by night
circiter	about	**orior, orīrī, ortus**	I arise
color, colōris, *m.*	colour	**coorior, coorīrī, coortus**	I arise;
cursus, -ūs, *m.*	course; speed; running		I break out
dēmōnstrō (1)	I point out, show	**adorior, adorīrī, adortus**	I arise, attack
extrēmus-a-um	outermost, furthest, last	**palūs, palūdis**, *f.*	marsh
		paulīsper	for a little
ferē	almost, about	**plērīque, plēraeque, plēraque**	most
ignārus-a-um	ignorant (of)		
ignōrō (1)	I am ignorant of	**prōtinus** (adverb)	straightway
incola, -ae, *c.*	inhabitant	**rēmus, -ī**, *m.*	oar
incolō, -ere, incoluī	I inhabit	**sustineō, -ēre, sustinuī, sustentum**	I sustain, withstand
māne	early (in the morning)		

90	**pecoris** of cattle.
95	**repertus** supply **est**.
98	**Cantium** = Kent. **utī** = **ut**.
99	**dē imprōvīsō** unexpectedly, by surprise.
100	**ēruptiōne** a sally.
102	**dētrīmentīs** losses.
105–6	**quid vectīgālis in annōs singulōs Britannia penderet** what (of) taxes Britain should pay each year . . . . Caesar speaks here as if Britain were now a province of the Roman Empire; this was distinctly premature.
109	**dēperierant** = **perierant. duōbus commeātibus** by two trips.
112	**dēsīderārētur** was missed = was lost. **secundā vigiliā:** about 10 p.m.

Word building

Make sure that you know the following compounds of
faciō, facere, fēcī, factum   I make; I do

afficiō, afficere, affēcī, affectum	I treat, affect
cōnficiō	I exhaust, accomplish
dēficiō	I fail; I run out
efficiō + ut	I effect; I achieve that
interficiō	I kill
patefaciō	I open
perficiō	I accomplish
praeficiō + acc. and dat.	I put (someone) in command of (someone or something)
prōficiō	I make progress, accomplish
reficiō	I repair, refresh
satisficiō + dat.	I satisfy
sufficiō	I supply, furnish

Give English derivatives from as many of these compounds as you can.

Gerundives of obligation

duae legiōnēs ad Britanniam trānsportandae sunt
Two legions are to be transported to Britain = Two legions must be transported to Britain.

Used in this way, as a complement to the verb **esse**, the gerundive expresses 'must', 'ought', 'have to' (obligation).

duae legiōnēs trānsportandae erant
Two legions had to be transported.

Caesar monendus erit
Caesar will have to be warned.

Exercise 3.1

Translate

1 castra in locō idōneō pōnenda sunt.
2 nāvēs in terram dēdūcendae sunt.
3 frūmentum cotīdiē in castra cōnferendum erat.
4 nāvēs tempestāte afflīctae reficiendae erunt.
5 nūntius statim ad ducem mittendus fuit.
6 The camp must be defended bravely.
7 The cavalry must be recalled at once.
8 The town had to be attacked from both sides.
9 The legions had to be taken back to Gaul.
10 The hostages will have to be guarded carefully.

The person who has to perform the action goes into the dative, e.g.
hoc mihi agendum est This is to be done by me = I must do this.

Exercise 3.2

Translate

1 castra Rōmānīs in locō idōneō pōnenda erant.
2 nāvēs mīlitibus in terram dēdūcendae erant.
3 frūmentum ūnī legiōnī cotīdiē cōnferendum est.
4 nāvēs tempestāte afflīctae nōbīs reficiendae erunt.
5 nūntius ad ducem lēgātō statim mittendus fuit.
6 omnia Caesarī eōdem tempore agenda erant.
7 quid nōbīs faciendum est?
8 hostēs vōbīs nōn longius persequendī sunt.
9 You must defend the camp bravely, soldiers.
10 Caesar had to recall the cavalry at once.
11 We had to attack the town from both sides.
12 The soldiers had to drive the enemy from the shore.
13 We must follow the standard.
14 The Britons will have to ask for peace.
15 You must send a letter to your father.

tibi fugiendum est = you must flee (literally: it is to be fled by you).
Intransitive verbs, i.e. verbs which cannot take an object, use the
neuter of the gerundive to express obligation.

Exercise 3.3

Translate

1 nōbīs festīnandūm est.
2 mīlitibus in flūctūs dēsiliendum erat.
3 tibi statim proficīscendum est.
4 puerīs domī manendum erat.
5 patrī ad urbem celeriter redeundum fuit.
6 We must not flee.
7 The soldiers had to fight in the waves.
8 You must go home at once.
9 We must not delay.
10 The legion had to stay in the camp to guard the ships.

Exercise 3.4

Quintus Cicero, the brother of the great Cicero, was on this expedition serving as a legionary commander. When Cicero heard that he had arrived safely in Britain, he wrote to him as follows (August 54 BC):

> ō iūcundās mihi tuās dē Britanniā litterās! timēbam Oceanum, timēbam lītus īnsulae; reliqua nōn contemnō, sed plūs habent spēī quam timōris, magisque sum sollicitātus exspectātiōne eā quam metū. tē vērō māteriam scrībendī ēgregiam habēre videō:
> 5 quōs tū sitūs, quās nātūrās rērum et locōrum, quās gentēs, quōs mōrēs, quem vērō ipsum imperātōrem habēs!

1	ō . . . litterās: exclamations usually go into the accusative.
3	sum sollicitātus I was worried. exspectātiōne eā by that waiting (for news).
4	vērō in fact, to be sure.

What makes Cicero most nervous on his brother's behalf? What does he seem to think is the greatest benefit his brother will gain from being on this expedition?

In April Cicero had written to Caesar, recommending a young lawyer called C. Trebatius, who hoped to gain something by being attached to Caesar's staff:

> Cicerō Caesarī Imperātōrī s.d.
> vidē quam mihi persuāserim tē mē esse alterum, nōn modo in eīs rēbus quae ad mē ipsum sed etiam in eīs quae ad meōs pertinent. mittō ad te Gāium Trebātium, dē quō tibi haec spondeō,
> 5 probiōrem hominem, meliōrem virum esse nēminem, quī familiam dūcit in iūre cīvīlī, vir singulārī memoriā, summā scientiā. huic nec tribūnātum neque praefectūram neque ūllius beneficiī certum nōmen petō; tōtum hominem tibi trādō. cūrā ut valeās et mē, ut amās, amā.

1	s.d. = salūtem dat.
2	tē mē esse alterum that you are my second self.
3	pertinent ad concern. meōs my friends.
4–5	spondeō I guarantee. probiōrem more upright.
6	familiam dūcit leads the field. in iūre cīvīlī in civil law.
	vir singulārī memoriā a man of uniquely good memory.
7–8	praefectūram prefectship. A praefectus could be put in charge of various jobs, military and civilian. ūllius beneficiī certum nōmen the specific name of any benefit = any particular piece of patronage.
9	mē, ut amās, amā love me, as you do.

Translate this letter. How do you think Caesar would have felt when he received it?

By the middle of June Cicero seems to have heard little from
Trebatius and wrote to him:

> ego tē commendāre nōn dēsistō, sed quid prōficiam ex tē scīre
> cupiō. spem maximam habeō in Balbō, ad quem dīligentissimē et
> saepissimē scrībō. illud soleō admīrārī, nōn mē totiēns accipere
> tuās litterās, quotiēns a Quīntō mihi frātre adferantur.
> 5   in Britanniā nihil esse audiō neque aurī neque argentī. id sī
> ita est, essedum aliquod capiās suādeō et ad nōs quam prīmum
> recurrās. sīn autem sine Britanniā adsequī quod volumus
> possumus, perfice ut sīs in familiāribus Caesaris. multum tē in eō
> frāter adiuvābit meus, multum Balbus, sed, mihi crēde, tuus
> 10   pudor et labor plūrimum. habēs imperātōrem līberālissimum,
> aetātem opportūnissimam, commendātiōnem certē singulārem,
> ut tibi ūnum timendum sit, ne ipse tibi dēfuisse videāris.

1	**prōficiam** I am achieving.
2	**Balbō** Balbus was Caesar's private secretary and very influential.
3–4	**totiēns ... quotiēns** so often ... as often = as often as ....
4	**tuās litterās** letters from you.
6	**capiās suādeō** I advise you to get.
7	**sīn** but if.
10	**pudor** modesty.
11	**certē** certainly. **singulārem** unique, exceptional.

Translate the first paragraph and answer the questions on the second

1 What does Cicero hear about Britain? What advice does he give
  Trebatius? Does this seem to you a serious suggestion?
2 1.7 **sine Britanniā**: explain what Cicero means.
3 1.7 **quod volumus**: what do they want?
4 What three things will help Trebatius to achieve his object? Which
  does Cicero consider the most important?
5 What three advantages has Trebatius got? Which does Cicero
  consider the most valuable?
6 What does Cicero say Trebatius should beware of?
7 What does the letter suggest about Trebatius's character and
  Cicero's attitude to him?

PS In fact Trebatius never followed Caesar to Britain when the
invasion took place the following month; he was nervous of the sea
crossing.

Exercise 3.5

Translate into Latin

When Caesar reached the river Thames, he saw that Cassivellaunus
had drawn up large forces on the other bank. He ordered the cavalry
to cross the river and attack the enemy and the legions to follow
quickly. The soldiers crossed the river with such speed and courage
that the enemy could not withstand their attack and fled. When
Caesar learnt where they had gone, he pursued them and took
Cassivellaunus's town. When Cassivellaunus learnt this, he despaired
and sent ambassadors to ask for peace.

Chapter IV

The revolt in Gaul

Caesar had returned from Britain to Gaul rather sooner than he had intended because of reports of unrest among the Gauls. His troops were hardly settled in their winter quarters when a major revolt broke out, in which Quintus Cicero was to prove the hero.

quod eō annō frūmentum in Galliā propter siccitātēs angustius
prōvēnerat, ipse quoad legiōnēs collocātās esse mūnītaque
hīberna cognōvisset, in Galliā morārī cōnstituit.

5   diēbus circiter quīndecim quibus in hīberna ventum est,
initium repentīnī tumultūs ac dēfectiōnis ortum est ā duōbus
prīncipibus Ambiorige et Catuvolcō, quī cum ad fīnēs rēgnī suī
Sabīnō Cottaeque praestō fuissent frūmentumque in hīberna
comportāvissent, subitō lignātōribus oppressīs magnā cum manū
ad castra oppugnanda vēnērunt.

10   cum celeriter nostrī arma cēpissent vāllumque ascendissent
atque, ūnā ex parte equitibus ēmissīs, equestrī proeliō superiōrēs
fuissent, dēspērātā rē hostēs ab oppugnātiōne suōs redūxērunt.
tum mōre suō conclāmāvērunt ut aliquis ex nostrīs ad colloquium
prōdīret.

15   mittitur ad eōs colloquendī causā C.Arpīnēius, eques
Rōmānus, et Q.Iūnius, quibus Ambiorix haec dīxit: tōtam
Galliam in armīs esse atque omnia hīberna ex commūnī cōnsiliō
simul oppugnārī; pollicētur tamen prō Caesaris in sē beneficiīs
tūtum iter sē Rōmānīs per fīnēs suōs datūrum esse, sī vellent ad
20   proxima hīberna contendere.

Sabīnus hīs verbīs dēceptus cēterīs persuādet ut hīs
condiciōnibus acceptīs ad hīberna Q.Cicerōnis contenderent,
quae quīnquāgintā mīlia passuum aberant.

Rōmānī prīmā lūce profectī ex castrīs, duo mīlia passuum
25   prōgressī in īnsidiās incidērunt ab hostibus in silvīs collocātās. ab
hostibus circumventī ā prīmā lūce ad hōram octāvam fortiter
resistēbant, multīs vulneribus acceptīs. tum Sabīnus, rē
dēspērātā, cum procul Ambiorigem suōs cohortantem
cōnspexisset, interpretem suum ad eum mittit ad rogandum ut
30   sibi mīlitibusque parcat. ille respondit, sī Sabīnus vellet sēcum
colloquī, licēre. Sabīnus tribūnōs mīlitum quōs circum sē
habēbat sē sequī iussit et, cum propius Ambiorigem accessisset,
iussus arma abicere imperātum facit suīsque ut idem faciant
imperat.

35   interim, dum dē condiciōnibus agunt, Sabīnus paulātim

40

lignātōrēs

1	**siccitātēs** drought.
1–2	**angustius prōvēnerat** had grown more poorly.
2	**quoad** until.
3	**in Galliā morārī:** Caesar used to return to Italy each year when the campaigning season was finished.
4	**diēbus quīndecim quibus . . . ventum est** within fifteen days after winter quarters were reached (literally: it was come to winter quarters).
7	**Sabīnō Cottaeque praestō fuissent** had been there to help Sabinus and Cotta. Sabinus and Cotta had one and a half legions in their camp, which was in the district of the Eburones, between the Rhine and the Moselle (see map, p. 14).
8	**lignātōribus** wood collectors = a party collecting wood for fuel.
12	**hostēs:** nominative case.
18	**prō Caesaris in sē beneficiīs** in return for the kindnesses Caesar had shown him.
21	Sabinus was strongly opposed by Cotta and others but he overruled them. He was completely deceived by Ambiorix and did not take adequate precautions on the march, according to Caesar. Cicero's legion was stationed in the territory of the Nervii, about fifty miles to the west.
29	**interpretem suum** his interpreter.
31	**licēre** it was allowed = he might.
33	**imperātum facit** carries out the order.
35	**dē condiciōnibus agunt: agere dē** to discuss.

circumventus interficitur. tum vērō suō mōre victōriam
conclāmant impetūque in nostrōs factō ordinēs perturbant.
L.Cotta pugnāns interficitur cum maximā parte mīlitum; reliquī
sē in castra recipiunt unde erant ēgressī. aegrē ad noctem
40 oppugnātiōnem sustinent. noctū ad ūnum omnēs dēspērātā
salūte sē ipsī interficiunt. paucī ex proeliō ēlāpsī incertīs
itineribus per silvās ad T.Labiēnum lēgātum in hīberna
perveniunt atque eum dē rēbus gestīs certiōrem faciunt.

hāc victōriā sublātī Gallī, nūntiīs ad gentēs fīnitimās missīs,
45 ad Cicerōnis hīberna advolant. nostrī celeriter ad arma
concurrunt, vāllum ascendunt; aegrē is diēs sustentātur.
mittuntur ad Caesarem cōnfestim litterae; obsessīs omnibus viīs
missī intercipiuntur. hostēs posterō diē multō māiōribus coāctīs
cōpiīs castra oppugnant, fossam complent. eādem ratiōne quā
50 prīdiē ā nostrīs resistitur. nūlla pars nocturnī temporis ad
labōrem intermittitur; nōn aegrīs, nōn vulnerātīs facultās quiētis
datur. ipse Cicerō, cum tenuissimā valētūdine esset, nē
nocturnum quidem sibi tempus ad quiētem relinquēbat, ut ultrō
mīlitēs concurrerent atque eum sibi parcere cōgerent.

55 tum ducēs Gallōrum colloquī sē velle dīcunt. factā
potestāte, eadem quae cum Sabīnō ēgerant, commemorant.
omnem esse in armīs Galliam: Germānōs Rhēnum trānsiisse;
Caesaris reliquōrumque hīberna oppugnārī; licēre illīs
incolumibus ex hībernīs discēdere et quāscumque in partēs velint
60 sine metū proficīscī. Cicerō ad haec ūnum modo respondit: nōn
esse cōnsuētūdinem populī Rōmānī accipere ab hoste armātō
condiciōnem: sī ab armīs discēdere velint, lēgātōs ad Caesarem
mittant; spērāre prō eius iūstitiā quae petiērint impetrātūrōs.

ab hāc spē repulsī Gallī vāllō pedum IX et fossā pedum XV
65 hīberna cingunt. septimō oppugnātiōnis diē, maximō coortō
ventō, fervefacta iacula in casās, quae mōre Gallicō strāmentīs
erant tēctae, iacere coepērunt, hae celeriter ignem
comprehendērunt et ventī magnitūdine in omnem locum
castrōrum distulērunt. hostēs, sīcutī partā iam victōriā, turrēs
70 testūdinēsque agere et scālīs vāllum ascendere coepērunt. at
tanta mīlitum virtūs fuit ut, cum ubīque flammā torrērentur
suaque omnia impedīmenta cōnflagrāre intellegerent, nēmō dē
vāllō dēscenderet, paene nē respiceret quidem quisquam, ac tum
omnēs ācerrimē fortissimēque pugnārent. hic diēs nostrīs longē
75 gravissimus fuit.

quantō erat in diēs gravior atque asperior oppugnātiō, tantō
crēbriōrēs litterae nūntiīque ad Caesarem mittēbantur. quōrum
pars dēprehēnsa in cōnspectū nostrōrum mīlitum cum cruciātū
necābātur. erat ūnus intus Nervius, nōmine Verticō, locō nātus
80 honestō, quī ā prīmā obsidiōne ad Cicerōnem perfūgerat
suamque eī fidem praestiterat. hic servō spē lībertātis magnīsque

36–7	**victōriam conclāmant** raise the victory cry.
40	**ad ūnum omnēs** all to a man.
42	**T. Labiēnum:** Labienus's camp was fifty miles or more south of Sabinus's in the territory of the Remi.
44	**sublātī** elated.
46	**aegrē is diēs sustentātur** that day is sustained with difficulty = they scarcely hold out that day.
48	**missī** the men sent = the messengers.
49	**fossam** the ditch. The **vāllum** consisted of a ditch and mound (**agger**); the earth dug out of the **fossa** formed the **agger**; the **agger** was surmounted by a palisade of stakes. **eādem ratiōne** in the same way.
50–1	**nūlla pars . . . intermittitur** no part of the night time is interrupted as far as work is concerned = work goes on night and day without interruption.
52	**cum** although. **tenuissimā valētūdine** in very poor health.
53	**ultrō** of their own accord.
56	**factā potestāte** permission having been granted.
63	**mittant** they should send. **spērāre . . . imperātūrōs (esse)** he (Cicero) expected that they would gain (i.e. be granted) . . . **prō eius iūstitiā** in view of his (Caesar's) sense of justice.
64	**vāllō pedum IX** with a rampart of nine feet = with a rampart nine feet high.
65	**cingunt** they surround.
66	**fervefacta iacula** burning spears = spears carrying fire. **casās** huts. **strāmentīs** with straw = thatch.
69	**sīcutī partā iam victōriā** as though they had now won the victory. **turrēs:** these were mobile towers, made to the height of the enemy fortifications.
70	**scālīs** with ladders. The Gauls had learnt the art of siege warfare partly from experience of Roman attacks and partly from Roman prisoners.
71–2	**cum . . . torrērentur** although they were being scorched. **sua omnia impedīmenta cōnflagrāre** that all their baggage was on fire.
73	**paene nē . . . quisquam** not even did anyone look back = almost no one even looked behind him.
76–7	**quantō gravior . . . tantō crēbriōrēs** the more serious . . . the more frequent . . .
78	**cum cruciātū** with crucifixion.
79	**intus:** i.e. inside the camp.
79–80	**locō nātus honestō** born in an honourable station = of honourable birth. This Gaul even owned slaves.
81	**praestiterat** had shown.

The enemy attack.

persuādet praemiīs ut litterās ad Caesarem dēferat. hās ille in
iaculō inligātās effert et Gallus inter Gallōs sine ūllā suspiciōne
versātus ad Caesarem pervēnit. ab eō dē perīculīs Cicerōnis
85 legiōnisque cognōscitur.

    Caesar acceptīs litterīs statim nūntiōs ad hīberna fīnitima
mittit; lēgātōs iubet cōnfestim proficīscī celeriterque ad sē
venīre. vēnit magnīs itineribus in fīnēs Nerviōrum. ibi ex captīvīs
cognōscit quae apud Cicerōnem gerantur quantōque in perīculō
90 rēs sit. tum cuidam ex equitibus Gallīs magnīs praemiīs
persuādet utī ad Cicerōnem epistolam dēferat. hanc Graecīs
cōnscrīptam litterīs mittit, nē interceptā epistolā nostrā ab
hostibus cōnsilia cognōscantur. sī adīre nōn possit, Caesar monet
eum ut trāgulam cum epistolā ad āmentum dēligātā intrā
95 mūnītiōnem castrōrum abiciat. in litterīs scrībit sē cum
legiōnibus profectum celeriter adfore; hortātur ut prīstinam
virtūtem retineat.

    Gallus, perīculum veritus, ut erat praeceptum, trāgulam
mittit. haec cāsū ad turrim adhaesit neque ā nostrīs bīduō
100 animadversa tertiō diē ā quōdam mīlite cōnspicitur, dēmpta ad
Cicerōnem dēfertur. ille perlēctam in conventū mīlitum recitat
maximāque omnēs laetitiā afficit. tum fūmī incendiōrum procul
vidēbantur, quae rēs omnem dubitātiōnem adventūs legiōnum
expulit.

105   Gallī, rē cognitā per explōrātōrēs, obsidiōnem relinquunt,
ad Caesarem omnibus cōpiīs contendunt. hae erant circiter mīlia
LX. Cicerō datā facultāte Gallum ab eōdem Verticōne quem
suprā dēmōnstrāvimus repetit, quī litterās ad Caesarem dēferat.
hunc admonet ut iter cautē dīligenterque faciat. perscrībit in
110 litterīs hostēs ab sē discessisse omnemque ad eum multitūdinem
convertisse. quibus litterīs circiter mediā nocte allātīs, Caesar
suōs certiōrēs facit eōsque ad dīmicandum animō cōnfirmat.

    posterō diē prīmā lūce movet castra et circiter mīlia
passuum quattuor prōgressus trāns vallem et rīvum
115 multitūdinem hostium cōnspicātur. erat magnī perīculī rēs
tantulīs cōpiīs inīquō locō dīmicāre; tum, quoniam obsidiōne
līberātum esse Cicerōnem sciēbat, aequō animō remittendum
esse dē celeritāte exīstimābat. cōnsēdit et quam aequissimō locō
potest castra commūnit, atque haec, etsī erant exigua per sē, vix
120 hominum mīlium septem, tamen angustiīs viārum quam maximē
potest contrahit, eō cōnsiliō ut in summam contemptiōnem
hostibus veniat.

    eō diē parvulīs equestribus proeliīs ad aquam factīs, utrīque
sēsē suō locō continent, Gallī, quod ampliōrēs cōpiās, quae
125 nōndum convēnerant, exspectābant; Caesar, sī forte timōris
simulātiōne hostēs in suum locum ēlicere posset, ut citrā vallem
prō castrīs contenderent.

82–3   **hās in iaculō inligātās** this letter tied on/in his javelin. It may have been wrapped round the shaft or, perhaps, inside the spear, if he removed the head and put it inside the shaft.

84   **versātus** moving.

85   **cognōscitur** it is learnt from him (impersonal use of passive).

87   **lēgātōs:** one of these was M. Crassus, son of the triumvir, who was killed next year with his father at Carrhae.

88   **magnīs itineribus** by forced marches.

90   **equitibus Gallīs:** these were Gauls serving in the Roman army.

94   **ad āmentum dēligātā** tied to the thong.

96   **prīstinam virtūtem** his previous courage = the courage he had shown so far.

98   **ut erat praeceptum** as it had been ordered = as he had been ordered (impersonal passive).

99   **adhaesit** stuck.   **bīduō** for two days.

100   **dēmpta** taken down.

102   **fūmī incendiōrum** smoke from the (watch) fires.

107–8 **Cicerō Gallum repetit, quī . . . dēferat** Cicero asks for the Gaul back, to carry . . . . The relative with the subjunctive expresses purpose.

112   **eōs ad dīmicandum animō cōnfirmat** strengthens them in mind for the fight.

114   **rīvum** a stream.

115   **magnī perīculī rēs** a matter of great danger = a very dangerous operation.

116   **tantulīs cōpiīs** with such small forces. The odds were ten to one against Caesar.

117   **aequō animō** with even mind = without worrying.

117–18 **remittendum esse dē celeritāte** he should reduce his speed.   **cōnsēdit** he halted.

119   **exigua per sē** tiny in itself = tiny in any case.

120   **angustiīs viārum** by the narrowness of the streets. Roman camps were laid out on a set pattern. There were four gates from which led streets, intersecting at Head Quarters; these were of regulation width, but to deceive the enemy Caesar on this occasion made them narrower than usual.

123   **ad aquam** at the water = at the stream.   **utrīque** both sides.

125   **sī forte** if by chance = in the hope that.

126   **ēlicere** to entice.   **citrā vallem** on his side of the valley.

aegrē	with difficulty, scarcely	condiciō, condiciōnis, *f*.	condition
angustus-a-um	narrow	consuētūdō, consuētūdinis, *f*.	custom
angustiae, -ārum, *f*.	narrowness; narrows, straits	etsī	even if, although
aequus-a-um	equal, level, fair; favourable	fīnitimus-a-um	neighbouring
		fossa, -ae, *f*.	ditch
inīquus-a-um	unequal, unfair, unfavourable	fūmus, -ī, *m*.	smoke
		ratiō, ratiōnis, *f*.	reason, way, method
beneficium, -ī, *n*.	benefit, kindness		
certus-a-um	certain, sure, reliable	repente	suddenly
incertus-a-um	uncertain	repentīnus-a-um	sudden
colloquor, -ī, collocūtus	I talk with	tumultus, -ūs, *m*.	upheaval, tumult
colloquium, -ī, *n*.	conversation, parley	vallēs, vallis, *f*.	valley
commūnis, -e	shared, common		

G Word building

Make sure that you know the following compounds of
capiō, capere, cēpī, captum

accipiō, accipere, accēpī, acceptum	I receive, accept	**percipiō**	I perceive
		praecipiō	I instruct, warn
dēcipiō	I deceive	**recipiō**	I take back
excipiō	I except	**mē recipiō**	I retreat
incipiō	I begin	**suscipiō**	I undertake

Give English derivatives from as many of these compounds as you can.

G Impersonal verbs

There are a few verbs in English which cannot be used with a personal subject, e.g. it rains, it thunders (you cannot say 'I rain'). Latin has rather more such verbs, and some of them are very common. For the moment learn

mihi licet (licēre, licuit) it is lawful for me, I am allowed, I may
mihi placet (placēre, placuit) it pleases me, I decide
mē oportet (oportēre, oportuit) it is right for me, I ought
mē pudet (pudēre, puduit) it shames me, I am ashamed

Note that after **licet** and **placet** the person goes into the dative; after **oportet** and **pudet** into the accusative;
thus: **Caesarī placuit**   Caesar decided
      Caesarem puduit   Caesar was ashamed

Exercise 4.1

Translate

1 sī vīs mēcum colloquī, tibi licet.
2 licet vōbīs ex hībernīs discēdere et sine metū proficīscī.
3 Sabīne, nōnne tē pudet hostium condiciōnēs accipere?
4 nōs oportet auxilium Cicerōnī statim ferre.
5 Caesarī placuit sine morā proficīscī.
6 tē oportuit iter celerius facere.
7 Gallīs placuit ad Cicerōnis castra advolāre.
8 If you wish, you may go now.
9 We ought to write a letter to mother.
10 Were you not ashamed to flee from the battle?
11 The general decided to lead his forces back to camp.
12 We were not allowed to stay in the city.

ꝺ The impersonal use of the passive voice

Intransitive verbs, i.e. verbs which cannot take a direct object,
cannot, either in English or Latin, be used personally in the passive,
e.g. it makes no sense to say: 'we were comed.'
But Latin does sometimes use intransitive verbs impersonally in the
passive, e.g.

quīndecim diēbus in hīberna ventum est Within fifteen days it was
come to winter quarters, i.e. winter quarters were reached within
fifteen days.
ab utrīsque ācriter pugnātum est It was fought fiercely by both sides,
i.e. both sides fought fiercely.
eādem ratiōne quā prīdiē ā nostrīs resistitur In the same way as the
day before it is resisted by our men, i.e. our men resist in the same
way as the day before.

Since we seldom use the passive like this in English, we often have to
use a personal subject and an active verb, as in the examples above,
when we translate.
Compare the use of gerundives:
nōbīs statim proficīscendum est It is to be set out by us at once = we
must set out at once.

Exercise 4.2

Translate

1 māne domō profectī sumus; merīdiē ad urbem ventum est.
2 proelium mox commissum est; diū et ācriter pugnātum est.
3 Gallī castra oppugnant; ā nostrīs ad arma celeriter concurritur.
4 Caesar legiōnēs ē castrīs ēdūxit; ventum est magnīs itineribus in fīnēs Nerviōrum.
5 Gallus, ut erat praeceptum, trāgulam mittit.
6 Gallī tōtum diem castra oppugnābant; ā nostrīs fortiter resistēbātur.
7 vōbīs hodiē domī manendum est.
8 tibi moriendum est.
9 nōbīs nunc est bibendum.
10 diū collocūtī sumus; tandem surgitur et domum reditum est.

An inscription on the base of a statue of the deified Julius Caesar

Exercise 4.3

The following passage concludes the story of the relief of Cicero's camp.
Without translating answer the questions below

    prīmā lūce hostium equitātus ad castra accēdit proeliumque cum
nostrīs equitibus committit. Caesar cōnsultō equitēs cēdere
sēque in castra recipere iubet; simul ex omnibus partibus castra
altiōre vāllō mūnīrī portāsque obstruī atque omnia haec cum
5 simulātiōne agī timōris iubet.
    quibus omnibus rēbus hostēs invītātī cōpiās trādūcunt
aciemque inīquō locō cōnstituunt; nostrīs vērō etiam dē vāllō
dēductīs, propius accēdunt et tēla intrā mūnītiōnem ex omnibus
partibus coniciunt, praecōnibusque circummissīs prōnūntiārī
10 iubent seu quis Gallus seu Rōmānus velit ante hōram tertiam ad
sē trānsīre, sine perīculō licēre; ac sīc nostrōs contempsērunt ut
aliī vāllum manū scindere, aliī fossās complēre incipiunt. tum

Caesar omnibus portīs ēruptiōne factā equitātūque ēmissō
celeriter hostēs in fugam dat, sīc utī pugnandī causā resisteret
15 nēmō, magnumque numerum ex eīs occīdit atque omnēs armīs
exuit.

longius prōsequī veritus, quod silvae palūdēsque
intercēdēbant, omnibus suīs incolumibus cōpiīs eōdem diē ad
Cicerōnem pervēnit. legiōne prōductā, cognōscit nōn decimum
20 quemque esse relictum mīlitem sine vulnere. ex eīs omnibus
iūdicat rēbus quantō cum perīculō et quantā cum virtūte rēs sint
administrātae. Cicerōnem prō eius meritō legiōnemque
collaudat; centuriōnēs singillātim tribūnōsque mīlitum appellat,
quōrum ēgregiam virtūtem testimōniō Cicerōnis cognōverat.

2	**cōnsultō** on purpose, by design.
4	**obstruī** to be blocked up.
6	**trādūcunt** lead across, i.e. across the stream onto the ground Caesar had chosen.
9	**praecōnibus** heralds.
10	**seu quis Gallus seu Rōmānus** if any Gaul or Roman.
12	**scindere** to tear apart.
16	**exuit** stripped.
19	**legiōne prōductā** the legion having been paraded.
19–20	**nōn decimum quemque** not one in ten.
22	**prō eius meritō** in accordance with his deserts.
23	**singillātim** individually.
24	**testimōniō** from the evidence.

Caesar legiōnem collaudat

1 When the Roman cavalry engaged the enemy, what orders did
Caesar give them? What did he order the soldiers to do inside the
camp?
2 What was the purpose of all this? How well did the plan succeed?
3 What made the Gauls come even closer to the camp?
4 How did the Gauls show their contempt for the Roman defenders?
5 How did Caesar surprise the Gauls and with what result?
6 Why did he not pursue them further?
7 How many men had been wounded in Cicero's legion (reckon
4,500 men to a legion)?
8 What was Caesar's judgement on how things had been managed?
Whom did he praise?
9 From the second paragraph give an example of: an ablative
absolute; a connecting relative; a consecutive clause; an
impersonal verb.
10 Considering this chapter as a whole, what qualities did (a) Cicero
(b) Caesar show in this operation?

PS In this revolt Caesar had lost more than a whole legion (Sabinus's command) and only saved Cicero by the skin of his teeth. He took savage reprisals on the Eburones, in whose territory the revolt had begun. But in 52 BC there was a far more serious rising which began in central Gaul, led by Vercingetorix. At Gergovia Caesar suffered his first defeat at the hands of the Gauls. The revolt only ended when he besieged Vercingetorix in the hill town of Alesia with a double encircling wall. Caesar beat off a Gallic relief force of 250,000 men and eventually Vercingetorix surrendered to save his men. Caesar then spent his final eighteen months in Gaul settling the province by moderate measures.

Exercise 4.4

Translate into Latin
After defeating and destroying the forces of Sabinus, the Gauls swooped down on Cicero's camp. The Romans ran to arms and mounted the rampart. The Gauls attacked so fiercely that the Romans could scarcely withstand their onslaught. All the messengers whom Cicero sent to Caesar were caught and killed by the enemy. At last Cicero sent a Gallic slave, who reached Caesar safely. Caesar sent a message to Cicero that he would bring help at once. Cicero's soldiers soon saw the smoke of Caesar's campfires and knew that he was not far off.
swoop down **advolāre**   camp fire **incendium**, -ī, *n*.

The Dying Gaul

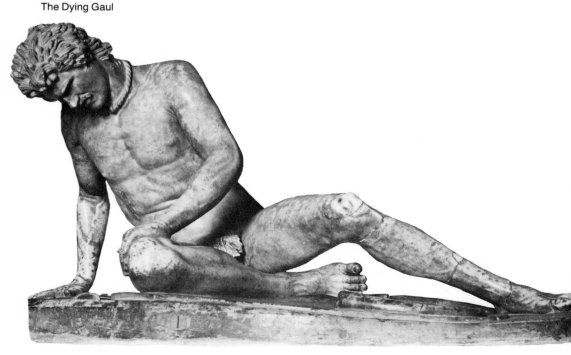

CICERO

Outline of Cicero's life

BC

106	Cicero born at Arpinum, 3 January
102	Quintus Cicero born
94?	Family moves to Rome for the boys' schooling
89	Cicero serves in the Social War
81	Cicero's first speech in the law courts
79–8	Cicero and Quintus travel in Greece and Asia
76	Cicero marries Terentia
75	Tullia born
75–4	Cicero quaestor in Sicily
70	Cicero aedile; speech aganst Verres
66	Cicero praetor
65	birth of Marcus
63	Cicero consul; Catiline's conspiracy
59	Caesar consul; First Triumvirate – Caesar, Pompey, Crassus
58	Cicero driven into exile by Clodius
57	Cicero recalled from exile
53	Battle of Carrhae – Crassus killed
52	Murder of Clodius by Milo
51	Cicero governor of Cilicia
50	Cicero returns to Italy (November)
49	Caesar crosses Rubicon (January) – Civil War. Cicero joins Pompey in Greece
48	Battle of Pharsalus – death of Pompey
47	Cicero returns to Italy; he is pardoned by Caesar
46	Cicero divorces Terentia
45	Death of Tullia
44	Assassination of Julius Caesar, 15 March. September, Cicero returns to Rome and attacks Antony in Senate; he leads the Senate against Antony
43	November, Second Triumvirate – Antony, Octavian, Lepidus; proscriptions. 9 December, Cicero murdered

CHAPTER V

The young Cicero

Arpīnum today

Marcus Tullius Cicerō Arpīnī nātus est, quod oppidum sexāgintā
mīlia passuum Rōmā abest in collibus salūbribus. hīc genus eius
semper habitāverat, māiōrēsque eius fuerant regiōnis prīncipēs,
domī nōbilēs, ut dīcunt. pater eius erat vir ērudītus, quī cum
5   īnfirmā esset valētūdine, hīc ferē aetātem ēgit in litterīs. Cicerō,
cum clārissimus in rēpūblicā factus esset multīsque
occupātiōnibus distinērētur, locum tamen nātālem etiam tum
dīligēbat revīsēbatque quotiēns licēbat. dē quō locō in dialogō
Dē Lēgibus haec dīcit.

10   Atticus: anteā mīrābar (nihil enim esse in hīs locīs nisi saxa et
montēs cōgitābam) tē tam valdē hōc locō dēlectārī; nunc contrā
mīror tē, cum Rōmā absīs, usquam potius esse.
    Cicerō: ego vērō, cum licet plūrēs diēs abesse, praesertim hōc
tempore annī, et amoenitātem et salūbritātem hanc sequor; rārō
15   autem licet. sed nīmīrum mē aliā quoque causā dēlectat, quae tē
nōn attingit, Tite.
    Atticus: quae tandem ista causa est?
    Cicerō: quia, sī vērum dīcimus, haec est mea et frātris meī
germāna patria; hīc enim ortī stirpe antīquissimā sumus, hīc
20   sacra, hīc genus, hīc māiōrum vestīgia. quid plūra? hanc vidēs
vīllam, ut nunc quidem est, lautius aedificātam patris nostrī
studiō, quī cum īnfirmā esset valētūdine, hīc ferē aetātem ēgit in
litterīs. sed hōc ipsō in locō, cum avus vīveret et antīquō mōre
parva esset vīlla, mē scītō esse nātum. quā rē inest nescioquid et
25   latet in animō ac sēnsū meō, quō plūs hic locus fortasse dēlectat,
sī quidem etiam ille sapientissimus vir, Ithacam ut vidēret,
immortālitātem scrībitur repudiāsse.

The sacra

2   **salūbribus** health giving.
4   **domī nōbilēs** noble at home = local nobility.   **cum** since.
5   **ferē aetātem** nearly all his life.
7   **distinērētur** was detained, occupied.
8   **quotiēns** as often as.
8–9   **in dialogō** *Dē Lēgibus:* in the dialogue *On the Laws*, Cicero visits
           Arpinum with his brother Quintus and his oldest friend, Atticus. In
           this passage they have just arrived there, Atticus for the first time.

14   **amoenitātem** beauty.
15   **nīmīrum** certainly.
16   **nōn attingit** does not concern.
17   **quae tandem** whatever . . . ?   **tandem** can be used to emphasize
           interrogative words.
19–20 **germāna patria** my true homeland.   **stirpe antīquissimā** from a very
           ancient stock, family.   **sacra** sacred things, i.e. family cults.
20   **māiōrum vestīgia** traces of our ancestors.   **quid plūra?** why (should I
           say) more?
21   **lautius** more grandly.
24–5   **scītō** know: the imperative of **sciō.**   **quā rē** for this reason.   **nescioquid
           . . . quō** something or other . . . because of which . . . .   **latet** is hidden
           = lies deep in.
26   **sī quidem** since, seeing that.
           ille sapientissimus vir: Odysseus, who in Homer's *Odyssey* was
           promised immortality by the nymph Calypso, if he would stay with her.
           But he preferred to return to his home, Ithaca.
27   **repudiāsse** to have rejected.

ūnum frātrem habēbat, nōmine Quīntum, quattuor annīs
nātū minōrem, quem valdē dīligēbat. prīmum pater fīliōs domī
30 ipse ēdūcāvit, deinde, cum Cicerō duodecim annōs nātus esset,
Rōmam eōs dūxit ut optimīs ā praeceptōribus ērudīrentur. ibi
Cicerō omnibus cēterīs discipulīs mox excellēbat, inter quōs erat
Titus Pompōnius Atticus, quōcum amīcitiam iniit quae tōtam per
vītam dūrātūra erat.

35 vix togam virīlem assūmpserat cum mīlitāre coāctus est in
bellō quod Rōmānī cum sociīs Italicīs suscēperant; sed eōdem
annō bellum fēlīciter cōnfectum est, ita ut Cicerōnī ad studia
redīre licēret. iam vērō nōn modo ingeniō excellēbat Cicerō, sed
etiam cupidissimus erat glōriae in rē pūblicā adipīscendae.
40 eōrum quī in rē pūblicā ēnitēbant plērīque aut in rē mīlitārī
excellēbant aut in rēbus forēnsibus atque urbānīs. Cicerōnī
scīlicet ipsī multō minus placēbat rēs mīlitāris quam studia
ēloquentiae et iūris cīvīlis. dēcrēvit igitur in forō et iūdiciīs
ēnitēre, quamquam bene sciēbat reī mīlitāris glōriam plūs
45 dignitātis adferre ad honōrēs adipīscendōs quam glōriam rērum
forēnsium. multīs post annīs cum causam ageret prō L. Mūrēnā,
virō mīlitārī, quem accūsābat Servius Sulpicius, vir iūris cīvīlīs
valdē perītus, haec dīxit:
quī potest dubitārī quīn ad cōnsulātum adipīscendum multō
50 plūs adferat dignitātis reī mīlitāris quam iūris cīvīlis glōria?
vigilās tū dē nocte ut tuīs cōnsultōribus respondeās, ille ut eō quō
intendit mātūrē cum exercitū perveniat; tē gallōrum, illum
būcinārum cantus exsuscitat; tū āctiōnem īnstituis, ille aciem
īnstruit; tū cavēs nē tuī cōnsultōrēs, ille nē urbēs aut castra
55 capiantur. ac nīmīrum – dīcendum est enim quod sentiō – reī
mīlitāris virtūs praestat cēterīs omnibus. haec nōmen populō
Rōmānō, haec huic urbī aeternam glōriam peperit, haec orbem
terrārum pārēre huic imperiō coēgit; omnēs urbānae rēs, omnia
haec nostra praeclāra studia et haec forēnsis laus et industria
60 latet in tūtēlā ac praesidiō bellicae virtūtis. simul atque increpuit
suspiciō tumultūs, artēs īlicō nostrae conticēscunt.
ad studia reversus Cicerō optimum quemque ōrātōrum
causās in iūdiciīs agentem audiēbat; iūriscōnsultīs, cum
cōnsultōribus respondērent, aderat; Molōnī Rhodiō, cum
65 Rōmam vēnisset, operam dedit. iuvenis erat summā industriā.
multīs post annīs cum causam prō iuvene Caeliō ageret, ad studia
sua iuvenīlia respiciēns iūdicēs rogāvit cūr tam paucī in arte
dīcendī excellere cōnārentur: 'an vōs aliam causam esse putātis
cūr in tantīs praemiīs ēloquentiae, tantā voluptāte dīcendī, tantā
70 laude, tantō honōre, tam sint paucī semperque fuerint quī in hōc
labōre versentur? obterendae sunt omnēs voluptātē,
relinquenda studia dēlectātiōnis, lūdus, iocus, convīvium, sermō
paene familiārium dēserendus.'

29	**nātū minōrem** lesser by birth = younger.
35	**togam virīlem:** the plain white toga, worn by adult citizens, was assumed by boys when they were sixteen or seventeen.
36	**bellō . . . cum sociīs Italicīs:** in this war the Italian allies (**sociī**) of Rome fought to win the citizenship. This was granted them in 89 B.C., which almost immediately ended the war.
38	**iam vērō** now (introducing a new point).
39	**adipīscendae: adipīscor, -ī** I obtain, win.
40	**ēnitēbant** shone = distinguished themselves. **in rē mīlitārī** in the military sphere, in military life.
41	**in rēbus forēnsibus atque urbānīs** in affairs of the forum and the city. The **forum** is used of public life in general and the law courts in particular; **urbs** nearly always means Rome. **scīlicet** of course.
43	**iūris cīvīlis** civil law.
44–5	**plūs dignitātis** more importance, more weight.
48	**perītus** skilled in (+ genitive).
49	**quī potest dubitārī quīn** how can it be doubted that . . .
51	**tū**: i.e. the lawyer. **ille**: i.e. the general. **cōnsultōribus** to your clients.
52	**mātūrē** in good time.
52–3	**gallōrum . . . cantus** the crowing of cocks. **būcinārum cantus** the blaring of trumpets. **āctiōnem īnstituis** start a legal action.
55	**nīmīrum** in fact.
55–6	**reī mīlitāris virtūs** excellence in the military sphere.
56	**praestat** is superior to.
57	**peperit** has won.
59	**haec forēnsis laus** praise won here in the forum.
60	**latet in tūtēlā ac praesidiō** lies hidden in the safety and protection = lies safe in the protection . . . .
60–1	**simul atque increpuit suspiciō tumultūs** as soon as there has sounded a suspicion (= a whisper) of disorder . . . .
61	**īlicō** at once. **conticēscunt** fall silent.
63	**iūriscōnsultīs** experts in the law.
65	**operam dedit** paid close attention to.
66	**prō Caeliō:** on Caelius, see pages 76 and 115.
67	**iuvenīlia** youthful. **iūdicēs** the jury. Law cases were heard before large juries; 75 would be a usual number.
69	**in tantīs praemiīs ēloquentiae** in such rewards for eloquence = when there are such great rewards for eloquence.
71	**versentur** take part in. **obterendae sunt** must be eradicated.
72	**studia dēlectātiōnis** the pursuit of amusement. **iocus** fun.
72–3	**sermō familiārium** conversation with one's friends.

Cicerō germānam suam patriam Atticō ostentat

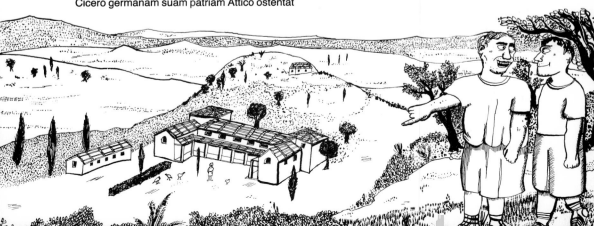

Cicerō tandem cōnfīdēbat sē adeō prōfēcisse ut initium in
75   forō dīcendī faceret. sex et vīgintī annōs nātus prō Sextō Rōsciō
causam tantā audāciā tantāque arte ēgit ut ab omnibus
laudārētur. sed posterō annō Rōmā discessit ut iter cum frātre in
Graeciam Asiamque faceret. cūr id fēcisset ipse exposuit:

'erat eō tempore in nōbīs summa gracilitās et īnfirmitās
80   corporis: quī habitus nōn procul abesse putātur ā vītae perīculō,
sī accēdit labor et laterum magna contentiō. . . . itaque cum mē et
amīcī et medicī hortārentur ut causās agere dēsisterem, quodvīs
potius perīculum mihi adeundum esse quam ā spērātā dīcendī
glōriā discēdendum putāvī. sed cum cēnsērem remissiōne et
85   moderātiōne vōcis et commūtātō genere dīcendī mē et perīculum
vītāre posse et temperātius dīcere, ea causa mihi in Asiam
proficīscendī fuit. itaque cum biennium essem versātus in causīs
et iam in forō celebrātum meum nōmen esset, Rōmā sum
profectus.'

90   Rōmam reversus uxōrem dūxit nōmine Terentiam,
mulierem nōbilī genere nātam, quācum multōs annōs contentus
vīxit. proximō annō fīliola, nōmine Tullia, eīs nāta est, quam
Cicerō semper valdē amābat. iam Cicerō ad eam aetātem
pervēnerat ubi cursum honōrum inīre licuit. quaestor ā populō
95   creātus in Siciliam missus est ubi rēs summā dīligentiā summāque
integritāte administrāvit.

Ⓥ

aeternus-a-um	eternal	**praesidium, -ī**, *n*.	protection; garrison
avus, -ī, *m*.	grandfather	**quidem**	indeed (emphasizes)
caveō, cavēre, cāvī, cautum	I beware	**sermō, sermōnis**, *m*.	conversation
cōgitō (1)	I think, reflect	**valeō** (2)	I am strong; I am well
dīligō, -ere, dīlēxī, dīlēctum	I love, am fond of	**validus-a-um**	strong; well
		valētūdō, valētūdinis, *f*.	health
fortasse	perhaps	**versor** (1) **in**	I take part in
genus, generis, *n*.	race, family; sort, kind	**vigilō** (1)	I stay awake
		vītō (1)	I avoid
īnfirmus-a-um	weak	**voluptās, voluptātis**, *f*.	pleasure
iste, ista, istud	this, that		
māiōrēs, -um, *m.pl.*	ancestors		

Ⓖ Word building

Make sure that you know the following compounds of eō, īre, i(v)ī, itum

abeō, abīre, abi(v)ī	I go away	**obeō**	I meet	**redeō**	I return
		pereō	I die	**subeō**	I undergo;
adeō	I approach	**praetereō**	I go past		come to mind
intereō	I die	**prōdeō**	I go forward		

74	**adeō prōfēcisse** had made such progress.
79	**in nōbīs = in mē:** Cicero often uses **nōs** for **ego** and **noster** for **meus.** **summa gracilitās** the greatest thinness; 'there was in me the greatest thinness . . . of body' = 'my body was extremely thin . . .'.
80	**quī habitus** this condition.
81	**sī accēdit . . . contentiō** if hard work and effort of the lungs is added to (**accēdit**) this = if besides this I worked too hard. **laterum contentiō** Cicero's early style of speaking was very violent and involved real physical effort. **cum** although.
82–3	**quodvīs potius perīculum . . . quam** any danger . . . rather than.
84–5	**remissiōne et moderātiōne vōcis** by resting and moderating my voice.
85	**commūtātō = mūtātō.**
86	**temperātius** with more restraint.
87	**biennium** for two years.
96	**integritāte** integrity.

Some common case usages

You have already met all of these in your reading.

1 The ablative of description

vir summā virtūte a man with the greatest courage (but English says: 'a man of the greatest courage.')
Marcus est magnō ingeniō Marcus is with great talent = Marcus is very clever.
Translate
vir summā hūmānitāte; mulier magnā prūdentiā; puer parvō ingeniō; puella mīrā pulchritūdine; iuvenis summā cōnstantiā; iste mīles est magnā crūdēlitāte; hic magister est summā gravitāte.

2 The ablative absolute

Caesare duce duae legiōnes Britannōs oppugnāvērunt
Caesar (being) leader two legions attacked the Britons.
The verb **esse** has no present participle; two nouns in the ablative case can form an ablative absolute.

Translate
1 Rōmulō rēge Rōma celeriter augēbātur.
2 Iūlius Caesar nātus est Mariō et Catulō cōnsulibus.
3 Aenēā duce Trōiānī tandem ad Italiam pervēnērunt.
4 Cicerōne quaestōre magna inopia erat frūmentī.
5 patre auctōre Cicerō, iuvenis īnfirmā valētūdine, iter in Asiam fēcit.

3 The partitive genitive

e.g. **multī cīvium; plērīque mīlitum; uterque nostrum.**

Latin frequently uses the partitive genitive after adjectives and pronouns and words such as **satis** (enough), **nimis** (too much), **parum** (too little).

What do the following phrases mean

plūs voluptātis; nihil malī; tantum ēloquentiae; satis cibī; nimis vīnī;
quid novī? aliquid bonī; parum virtūtis.

4 The predicative dative

In Caesar (page 20) we met: **haec rēs magnō ūsuī nostrīs fuit.**
What did this mean?

With certain nouns Latin uses a dative where we would expect a
complement in the nominative, e.g.
haec clādēs legiōnī magnō dēdecorī fuit
This disaster was (for) a great disgrace to the legion.
Amongst the commonest nouns used in this way are:

auxilium help	**dēdecus** disgrace	**impedīmentum** hindrance	**ūsus** use
cūra care, anxiety	**exitium** destruction	**praesidium** protection	

Exercise 5.1

Translate

1 Caesar duās legiōnēs in castrīs relīquit quae nāvibus praesidiō essent.
2 omnia dēsunt quae ad nāvēs reficiendās ūsuī sunt.
3 aquae altitūdō mīlitibus magnō impedīmentō erat.
4 patris valētūdō Cicerōnī magnae cūrae erat.
5 Atticī adventus Cicerōnī magnō auxiliō fuit.
6 iste canis domuī praesidiō est.
7 hoc scelus tibi dēdecorī est.
8 tempestās nāvī exitiō fuit.

Exercise 5.2

Translate

1 plērīque mīlitum Caesarem, ducem summā virtūte, dīligēbant.
2 eō duce legiōnēs nihil malī patiēbantur.
3 omnia dēsunt quae ad fluvium trānseundum ūsuī sunt; quid cōnsiliī prōpōnis?
4 Cicerō, vir ēgregiō ingeniō, tantum ēloquentiae praebēbat ut in
   causīs agendīs ēnitēret.
5 ēloquentia eī magnō auxiliō fuit in honōribus petendīs.
6 ad cōnsulātum adipīscendum plūs dignitātis adfert reī mīlitāris
   quam iūris cīvīlis glōria.
7 frātris meī perīculum mihi magnae cūrae est. audīvistīne aliquid novī?
8 Marce, nōlī plūs vīnī sūmere; iam nimis vīnī bibistī; timeō nē vīnum tibi exitiō sit.

Exercise 5.3

Translate the second paragraph; answer the questions on the last
   Cicerō, cum ē prōvinciā dēcēdēns in Italiam redīret, exīstimābat
   omnēs hominēs Rōmae suam quaestūram laudāre; sed longē

errāvit, ut dīxit ipse iūdicibus in causā quam multīs post annīs
prō Plancīō ēgit:

5   nōn vereor nē mihi aliquid, iūdicēs, videar adrogāre, sī dē
quaestūrā meā dīxerō. quamvīs enim illa flōruerit, tamen eum
mē posteā fuisse in maximīs imperiīs arbitror ut nōn ita multum
mihi glōriae sit ex quaestūrae laude repetendum. sed tamen nōn
vereor nē quis audeat dīcere ūllius in Siciliā quaestūram aut
10   clāriōrem aut grātiōrem fuisse. vērē mē Herculē hoc dīcam: sīc
tum exīstimābam, nihil aliud hominēs Rōmae nisi dē quaestūrā
meā loquī. omnibus eram vīsus in omnī officiō dīligentissimus;
excōgitātī quidem erant ā Siculīs honōrēs in mē inaudītī. itaque
hāc spē dēcēdēbam, ut mihi populum Rōmānum ultrō omnia
15   dēlātūrum esse putārem.
       at ego cum cāsū diēbus eīs iter faciendī causā dēcēdēns ē
prōvinciā Puteolōs forte vēnissem, cum plūrimī et lautissimī in
eīs locīs solent esse, concidī paene, iūdicēs, cum ex mē quīdam
quaesiisset quō diē Rōmā exiissem et num quid esset novī. cui
20   cum respondissem mē ē prōvinciā dēcēdere, 'etiam mē Herculē,'
inquit, 'ut opīnor, ex Africā.' huic ego iam stomachāns
fastīdiōsē, 'immō ex Siciliā,' inquam. tum quīdam, quasi quī
omnia scīret, 'quid? tu nescīs' inquit 'hunc quaestōrem Syrācūsīs
fuisse?' quid multa? dēstitī stomachārī et mē ūnum ex eīs fēcī quī
25   ad aquās vēnissent.

1	**dēcēdēns** retiring from.
5	**mihi aliquid adrogāre** to claim something for myself = to boast.
6	**quamvīs illa flōruerit** however it (my quaestorship) flourished = however successful it may have been.
6–7	**eum . . . fuisse** that I have been such a man, i.e. so successful. Cicero often uses parts of the demonstrative **is, ea, id** to look forward to a consequence clause.
10	**grātiōrem** more popular. **mē Herculē** by Hercules.
13	**excōgitātī erant** had been thought out, devised. **inaudītī** unheard of, unprecedented.
14	**ultrō** of their own accord, i.e. unasked.
15	**dēlātūrum** would offer.
17	**Puteolōs:** Puteoli was the most important port on the west coast of Italy and a fashionable seaside resort. **lautissimī** the smartest.
18	**concidī** I collapsed.
20	**etiam mē Herculē** yes, of course.
21	**stomachāns** getting annoyed.
22	**fastīdiōsē** scornfully. **immō** no.
22–3	**quasi quī omnia scīret** like a man who knew everything, a know-all.
23	**Syrācūsīs:** there were two quaestors in Sicily, one at the capital, Syracuse, the other in the west of the island at Lilybaeum. Cicero had been at Lilybaeum.
24–5	**quī ad aquās vēnissent** who had come to the waters, i.e. to bathe and enjoy themselves.

1 When Cicero arrived at Puteoli, whom did he find there?
2 **concidī paene** (1.18): why did he do this?
3 What did his questioner say when Cicero told him that he was returning from his province?
4 **stomachāns** (1.21): why was Cicero annoyed?
5 How did the second speaker get it wrong? How did Cicero react to his statement?
6 Who is Cicero laughing at in this passage? Do you think it would have amused the jury?
7 Cicero goes on to say: 'This experience probably did me more good than if everyone had congratulated me.' What lesson do you suppose he learnt from what happened?

The view across the bay of Puteoli

Exercise 5.4

Translate into Latin
1 Cicero led his friend to see his birthplace.
2 He told him why he loved the place so greatly.
3 He said that his family had lived there for very many years.
4 His father, who was of weak health, had spent nearly all his life there.
5 Although Cicero was seldom at lesiure, he loved the place so much that he often revisited it.

Exercise 5.5

Translate into Latin
When I had been made quaestor, I was sent to Sicilia. There I
performed all my duties so carefully that I thought that everyone in
Rome was praising me. But while I was returning from my province, I
went to visit a friend at Puteoli. There I met someone who did not
know where I had been or what I had done. I was very upset, but I
soon ceased to be indignant. If you leave Rome, everyone will soon
forget you.
or what **quidve**

The Roman Calendar

We have glossed dates when they occur in the text but they are so
common in Cicero's letters that it is worth explaining how they
worked.

Year dates

These are given by the names of the consuls, e.g. **C.Iūliō Caesare,
M.Calpurniō Bibulō cōnsulibus** (often abbreviated to **coss**) When
Caesar and Bibulus were consuls = 59 BC.

Dates of the month

Our names for the months are derived from the Roman names (the
Roman year used to begin in March, so that the seventh month was
September etc.):
Iānuārius, Februārius, Martius, Aprīlis, Māius, Iūnius, Quīntīlis
(later changed to Iūlius), Sextīlis (later changed to Augustus),
September, Octōber, November, December.
    The dates of the month were calculated from three fixed days:
Kalendae = 1st, **Nōnae** = 5th, **Idūs** = 13th, but in March, July
(Quīntīlis), October, May, the Nones are on the 7th and the Ides on
the 15th day.

Thus: **Kalendīs Aprīlibus**   1 April
      Nōnīs Sextīlibus   5 August
      Idibus Martiīs   15 March
Other dates are calculated by counting backwards from these fixed
days:
prīdiē Kalendās Sextīlēs the day before 1 August = 31 July.
ante diem tertium Idūs Martiās the third day before Ides of March =
13 March (counting is inclusive, i.e. count Ides (15th), 14th, 13th).

Dates are usually given in abbreviated form:
a.d. V Nōn. Oct. = **ante diem quīntum Nōnās Octōbrēs**
(Nones on 7th day) = 3 October.

CHAPTER VI

Cicero's consulship and exile

Cicerō cum aedīlis esset causam maximī mōmentī suscēpit;
C.Verrem enim de rēbus repetundīs accūsāvit, quī Siciliae
praetor prōvinciam per triennium ita vexāvit ac perdidit ut ea
restituī in antīquum statum nūllō modō posset. iam Siculīs
5   petentibus Cicerō nōmen eius dētulit. quam causam tantā vī
tantōque ingeniō ēgit ut Verrēs rē dēspērātā in exsilium fūgerit
antequam iūdicēs sententiās ferrent. quō ex tempore Cicerō
prīnceps habēbātur omnium quī causās agēbant.
       eōdem annō Quīntus frāter Pompōniam dūxit, Atticī
10 sorōrem; Pompōnia mulier erat difficilis ac superba, Quīntus
autem ad īrācundiam prōpēnsior, ita ut nōnnumquam in rīxās
inīre solērent. Cicerō ambōs dīligēbat et semper contendēbat ut
alterum alterī reconciliāret. brevī tempore sīc ad Atticum scrībit:

       Cicerō Atticō sal.
15   quod ad mē scrībis dē sorōre tuā, testis erit tibi ipsa quantae
mihi cūrae fuerit ut Quīntī frātris animus in eam esset is quī esse
dēbēret. quem cum esse offēnsiōrem arbitrārer, eās litterās ad
eum mīsī quibus et placārem ut frātrem et monērem ut minōrem
et obiūrgārem ut errantem. itaque ex eīs quae posteā saepe ab eō
20 ad mē scrīpta sunt cōnfīdō ita esse omnia ut et oporteat et
velīmus.
       Epīrōticam ēmptiōnem gaudeō tibi placēre. Quīntum
frātrem cotīdiē exspectāmus. Terentia magnōs articulōrum
dolōrēs habet. et tē et sorōrem tuam et mātrem maximē dīligit;
25 salūtemque plūrimam adscrībit et Tulliola, dēliciae nostrae. cūrā
ut valeās et nōs amēs et tibi persuādeās tē ā mē frāternē amārī.

       iam annus aderat cum Cicerōnī cōnsulātum petere licuit.
petītiōnem mātūrē iniit; sciēbat enim rārissimē ad cōnsulātum
pervēnisse novōs hominēs et nōbilēs honōrī suō vehementer
30 adversātūrōs esse. spērābat Atticum, quī multōrum nōbilium
familiāris erat, sē in petītiōne adiūtūrum esse; ad eum sīc scrībit:

       Cicerō Atticō salūtem dat:
L.Iūliō Caesare C.Marciō Figulō cōnsulibus, fīliolō mē auctum
esse scītō, salvā Terentiā.
35   abs tē iam diū nihil litterārum! ego dē meīs ad tē ratiōnibus
scrīpsī anteā dīligenter. hōc tempore Catilīnam, competītōrem
nostrum, dēfendere cōgitāmus. iūdicēs habēmus quōs volumus,
summā accūsātōris voluntāte. spērō, sī absolūtus erit,

1	**maximī mōmentī** of the greatest importance.

1   **maximī mōmentī** of the greatest importance.

2   **dē rēbus repetundīs accūsāvit** accused of extortion.

2–3   **Siciliae praetor** as governor of Sicily.

5   **nōmen eius dētulit** reported his name = prosecuted. There were no state prosecutors under Roman law; all actions were brought by private individuals. As Verres was a noble and had powerful supporters, Cicero was taking a big risk.

7   **sententiās ferrent** could cast their votes, i.e. deliver their verdict.

9   **dūxit** married.

11   **ad īrācundiam prōpēnsior** rather inclined to quick temper = rather irritable.

13   Quintus and Pomponia were married in 70 B.C. This is the earliest letter of Cicero which survives; it was written in November 68 B.C.

14   **sal.** = **salūtem dat.**

15   **quod ad mē scrībis** as for what you write . . .

16   **animus** feelings.　**esset is quī** should be such as . . .

17   **offēnsiōrem** rather offensive. We do not know what the cause of the quarrel was.

17–18   **eās litterās . . . quibus et placārem ut frātrem** such a letter by which I might both calm him as a brother . . . . The subjunctive expresses purpose; we would say 'a letter both to calm . . .'.

19   **obiūrgārem ut errantem** scold him as a wrong doer.

21   **velīmus** we should like.

22   **Epīrōticam ēmptiōnem** your Epirot purchase. Atticus had just bought an estate in Epirus (north-west Greece), where he was to spend much of his time from now on.

23   **exspectāmus:** Quintus was expected home from abroad.

23–4   **magnōs articulōrum dolōrēs** great pains of the joints = rheumatism.

25   **Tulliola, dēliciae nostrae** little Tullia, our darling.

26   **frāternē** like a brother.

27   **cōnsulātum petere licuit:** Cicero could stand for the consulship in 64 B.C. (you could not become consul until you were forty-three). The elections took place in July and Cicero began his election campaign in July 65 B.C., when he wrote this letter.

28   **petītiō** election campaign.

29   **novōs hominēs** new men. These were men whose ancestors had not held the consulship. In the last 100 years of the republic only four of the 200 consuls were new men.

33   **L. Iūliō . . . cōnsulibus:** Cicero's son, Marcus, was born the day Caesar (the great Caesar's uncle) and Figulus were elected consuls. Cicero announces his birth as if it were an event of national importance.

33–4   **fīliolō mē auctum esse** I have been blessed by a little son.

34   **scītō** know: the imperative of **sciō**.　**salvā Terentiā** Terentia being safe (ablative absolute).

35   **abs tē** = **ā tē** from you.　**dē meīs ratiōnibus** about my plans.

36   **competītōrem** fellow candidate, rival. Catiline was being prosecuted for extortion; Cicero did not in the end defend him.

37   **iūdicēs:** juries were large, e.g. seventy-five was a common number. Bribery was rife; you could tell how a case was likely to go when you knew who was to be on the jury. In this case Cicero evidently thought that the prosecutor would cooperate in securing Catiline's acquittal (**summā accūsātōris voluntāte**).

coniūnctiōrem illum nōbīs fore in ratiōne petītiōnis; sīn aliter
40 acciderit, hūmāniter ferēmus.

tuō adventū nōbīs opus est mātūrō; nam prōrsus summa
hominum opīniō est tuōs familiārēs nōbilēs hominēs honōrī
nostrō adversāriōs fore; ad eōrum voluntātem mihi conciliandam
maximō tē mihi ūsuī fore videō. quā rē Iānuāriō ineunte, ut
45 cōnstituistī, cūrā ut Rōmae sīs.

proximō annō Cicerō cōnsul creātus est, nōbilibus nōn
adversantibus, quod Catilīnae furōrem timēbant. ille iam bis in
comitiīs repulsus Cicerōne cōnsule coniūrātiōnem nefāriam iniit;
omnibus hominibus improbīs, omnibus egēnīs collēctīs,
50 mōliēbātur urbem incendere, cōnsulēs interficere, rempūblicam
ēvertere. Cicerō autem omnia eius cōnsilia cognōverat; nam
index quaedam, amīca coniūrātī, omnia quae Catilīna agēbat eī
dētulit. tandem testimōniō certō cōnfīsus Cicerō Catilīnam in
senātū vehementer accūsāvit. ille ex urbe ēlāpsus ad sociōs fūgit,
55 quī exercitum in Etrūriā comparābant. Cicerō coniūrātōs quī
Rōmae relictī erant īnsidiīs captōs in vincula iēcit. senātū
convocātō rettulit quid dē eīs quī in custōdiā tenēbantur fierī
placēret. plērīque senātōrum cēnsuērunt captīvōs statim
interficiendōs esse; Caesar tamen sententiam rogātus dīxit eōs in
60 carcere tenendōs esse. sed tandem senātus dēcrēvit ut
supplicium dē eīs sūmerētur, quod Cicerō prōtinus effēcit.

sīc Cicerō et urbe et rēpūblicā servātā ab omnibus est
laudātus, parēns patriae salūtātus. posteā tamen poenās datūrus
erat quod cīvēs sine iūdiciō ad supplicium trādidisset. quattuor
65 enim post annīs inimīcus eius nōmine Clōdius tribūnus plēbis
factus lēgem tulit ut sī quis cīvēs Rōmānōs sine iūdiciō
interfēcisset, eī aquā et igne interdīcerētur. Cicerō aliquamdiū
cunctātus tandem rē dēspērātā in exsilium fūgit.

in Graeciam profectus Thessalonīcae manēbat, animō
70 frāctus; identidem suam timiditātem dēplōrābat quod Clōdiī
furōrī nōn restitisset.

ad Terentiam sīc scrībit:

Tullius salūtem dat Terentiae suae et Tulliolae et Cicerōnī suīs.

nōlī putāre mē ad quemquam longiōrēs epistulās scrībere,
75 nisi sī quis ad mē plūra scrīpsit, cui putō rescrībī oportēre. neque
enim habeō quid scrībam neque hōc tempore quicquam
difficilius faciō. ad tē vērō et ad nostram Tulliolam nōn queō sine
plūrimīs lacrimīs scrībere. vōs enim videō esse miserrimās, quās
ego beātissimās semper esse voluī idque praestāre dēbuī, et, nisi
80 tam timidī fuissemus, praestitissem.

39	**coniūnctiōrem** more united (with me), i.e. they would cooperate to keep the third candidate out.
39–40	**sīn aliter acciderit** if it turns out otherwise, i.e. if Catiline is condemned.
40	**hūmāniter** in a civilized way.
41	**prōrsus** certainly.
41–2	**summa opīniō** the general opinion.
43	**nostrō = meō.**
44	**quā rē** and so.
47–8	**bis in comitiīs repulsus** having been twice rejected at the elections. Catiline stood for the consulship again in 63 B.C.
49	**egēnīs** needy.
50	**mōliēbātur** was plotting.

Cicerō Catilīnam in
senātū vehementer
accūsāvit

52	**index quaedam** a (female) informer.
57	**rettulit** put the question. At a debate in the Senate the presiding consul put the question forward. Senators were then asked for their opinion in order of seniority.
59	**sententiam rogātus** asked for his opinion.
61	**supplicium . . . sūmerētur** the death penalty should be exacted.
67	**eī aquā et igne interdīcerētur** he should be forbidden fire and water. This was an old legal formula, meaning 'should be banished' (**interdīcerētur** is impersonal passive: it should be forbidden to him.)
69	**Thessalonīcae:** Thessalonica is in north-east Greece.
73	**Cicerōnī:** i.e. his son, Marcus.
74	**longiōrēs epistulās:** Terentia had evidently complained that he was writing longer letters to other people than to her.
75–6	**neque habeō quid scrībam** I haven't anything to write.
77	**non queō** I cannot. **vērō** in fact.
79	**id praestāre dēbuī** I should have effected this, i.e. I should have made you happy.
79–80	**nisi fuissēmus . . . praestitissem** if I had not been . . . , I would have effected it.

longius, quoniam ita vōbīs placet, non discēdam; sed velim quam saepissimē litterās mittātis, praesertim sī quid est firmius quō spērēmus.

valēte, mea dēsīderia, valēte.

85 data ante diem tertium Nōnās Octōbrēs. Thessalonīcae.

tandem, quamquam Clōdius summā vī resistēbat, auctōre Pompēiō lēx lāta est quā Cicerō Rōmam revocātus est. reditum suum in litterīs ad Atticum scrīptīs sīc memorat:

nunc, etsī omnia aut scrīpta esse ā tuīs arbitror aut etiam
nūntiīs ac rūmore perlāta, tamen ea scrībam brevī quae tē putō potissimum ex meīs litterīs velle cognōscere. prīdiē Nōnās Sextīlēs Dyrrachiō sum profectus eō ipsō diē quō lēx est lāta dē nōbīs. Brundisium vēnī Nōnīs Sextīlibus. ibi mihi Tulliola mea fuit praestō nātālī suō ipsō diē, quī cāsū īdem nātālis erat
Brundisīnae colōniae; quae rēs animadversa ā multitūdine summā Brundisīnōrum grātulātiōne celebrāta est.

inde iter fēcī ita ut undique ad mē cum grātulātiōne lēgātī convēnerint. cum vēnissem ad portam Capēnam, gradūs templōrum ab īnfimō plēbe complētī erant. ā quā plausū maximō
cum esset mihi grātulātiō significāta, similis et frequentia et plausus mē usque ad Capitōlium celebrāvit in forōque et in ipsō Capitōliō mīranda multitūdō fuit.

postrīdiē in senātū senātuī grātiās ēgimus.

A fifteenth-century manuscript of Cicero's letter to Atticus

81	**longius:** Cicero had evidently thought of moving further from Rome than Thessalonica.
81–2	**velim . . . mittātis** I would like you to send . . .
82	**sī quid est firmius** if there is anything firmer by which we may hope = if there are any firmer grounds for hope.
84	**mea dēsīderia** my darlings.
90	**brevī** in brief.
90–1	**tē potissimum . . . velle** you would most like. **ex meīs litterīs** from a letter from me.
91–2	**prīdiē Nōnās Sextīlēs** on 4 August.
92	**Dyrrachiō:** he had moved from Thessalonica to Dyrrachium, a port in north-west Greece (modern Durazzo), so that he would be ready to cross to Italy as soon as he knew that the law about his recall had been passed.
94	**fuit praestō** was there. **praestō** is an adverb meaning present, at hand. **quī īdem nātālis** which was also (**īdem** the same) the birthday of the colony, i.e. the day on which the foundation of the colony was celebrated each year.
97	**lēgātī** ambassadors = deputations.
98	**porta Capēna:** this was the main gate of Rome leading to the south.
99	**ab īnfimō** from the bottom = from top to bottom.
100	**frequentia** crowds.
101	**usque ad** right up to.

cum vēnissem ad portam Capēnam

V

adversus-a-um	facing, opposite	**opīniō, opīniōnis**, *f.*	opinion
adversor (1)	I oppose	**obviam eō** + dative	I go to meet
adversārius-a-um	opposed to; an opponent	**poena, -ae**, *f.*	penalty, punishment
		poenās dō	pay the penalty
beātus-a-um	blessed, happy	**properō** (1)	I hasten
carcer, carceris, *m.*	prison	**rārus-a-um**	rare
cēnseō (2)	I vote, decide, think	**rārō** adverb	rarely
dēferō, dēferre,	I carry down; I offer;	**sententia, -ae**, *f.*	opinion, vote
dētulī, dēlātum	I report	**testis, testis**, *c.*	witness
nōmen dēferō	I indict, prosecute	**testimōnium, -ī**, *n.*	evidence
familiāris-e	friendly; a friend	**timidus-a-um**	timid, fearful
furor, furōris, *m.*	madness, fury	**timiditās,**	timidity
improbus-a-um	wicked, unprincipled	**timiditātis**, *f.*	
lapis, lapidis, *m.*	stone	**vinculum, -ī**, *n.*	chain
mātūrus-a-um	early, in good time		

G Word building

Make sure that you know the following compounds of
ferō, **ferre**, **tulī**, **lātum**

afferō, afferre, attulī, allātum	I bring to, report
anteferō	I prefer
auferō, auferre, abstulī, ablātum	I take away
cōnferō, cōnferre, contulī, collātum	I collect, compare
differō, differre, distulī, dīlātum	I postpone
efferō, efferre, extulī, ēlātum	I carry away
īnferō	I carry in
bellum īnferō	I attack
offerō, offerre, obtulī, oblātum	I offer
perferō	I endure
prōferō	I carry forward
referō, referre, rettulī, relātum	I report, carry back
pedem referō	I retreat

G 'Quis, quid' (indefinite pronoun)

You have long known the interrogative pronoun **quis**? = who?
quid? = what?

quis, quid is also an indefinite pronoun, used after **sī, nisi, nē, num**,
meaning 'anyone', 'anything', e.g.

sī quis cīvem interfēcerit, in exsilium expellētur
If anyone kills a citizen, he will be driven into exile.

litterās ad mē mitte, sī quid novī est
Send me a letter, if there is any news (literally 'anything of new').

Exercise 6.1

Translate

1 domum properāvimus nē quis nōs vidēret.
2 nesciēbāmus num quis nōbīs adversātūrus esset.
3 sī quis nōs oppugnāverit, eī fortiter resistēmus.
4 tibi properandum est nē quid calamitātis absēns accipiās.
5 Caesar, sī quem mīlitum labōrantem vīderat, auxilium mīsit.
6 auxilium tibi mīsī nē quid malī paterēris.

❭ The relative with the subjunctive

What is the difference between the following in grammar and meaning?
(a) Caesar cohortēs decem ad mare relīquit quae nāvibus praesidiō erant.
(b) Caesar cohortēs decem ad mare relīquit quae nāvibus praesidiō essent.

The relative pronoun is often used with the subjunctive to express purpose, e.g.
equitēs ā Quīntō Hirtiō ad Caesarem vēnērunt quī nūntiārent. . .
Cavalry came from Quintus Hirtius to Caesar who might announce. . .
i.e. to announce.

Exercise 6.2

Translate

1 Caesar legiōnem ēmīsit quae frūmentum colligeret.
2 Caesar locum idōneum cēpit in quō castra pōneret.
3 Quīntus Cicerō servum ē castrīs ēmīsit quī litterās ad Caesarem ferret.
4 magister librum puerō dedit quem legeret.
5 aliquid cibī inveniēmus quod edāmus.
6 servus nōbīs obviam iit quī dīceret quid accidisset.
7 cum Cicerō in Italiam rediisset, plūrimī convēnērunt quī eī grātulārentur.
8 dīcam quod vērum est, ex quō magnitūdinem miseriārum meārum perspicere possīs.

Exercise 6.3

Answer the questions below, without translating except where you are asked to

      sīc Cicerō summā cum glōriā, omnibus paene cīvibus
      gaudentibus, Rōmam rediit. inimīcī tamen eius adhūc furēbant,
      in prīmīs Clōdius, quī iam operās suās in viīs urbis palam
      incitābat. cum Cicerō exsul abesset, Clōdius domum eius in

Mōns Palātīnus

5 monte Palātīnō sitam dēlēverat; iam ex senātūs cōnsultō
   reficiēbātur. mēnse Novembrī sīc ad Atticum scrībit:

       armātīs hominibus ante diem tertium Nōnās Novembrēs
   expulsī sunt fabrī de āreā nostrā, disturbāta porticus Catulī, quae
   ex senātūs cōnsultō reficiēbātur et ad tēctum paene pervēnerat,
10 Quīntī frātris domus prīmō frācta coniectū lapidum ex āreā
   nostrā, deinde īnflammāta iussū Clōdiī, īnspectante urbe
   coniectīs ignibus, magnā querēlā et gemitū omnium hominum.
   ille vel ante dēmēns ruere, post hunc furōrem nihil nisi caedem
   inimīcōrum cōgitāre, servīs apertē spem lībertātis ostendere.
15   itaque ante diem tertium Idūs Novembrēs, cum Sacrā Viā
   dēscenderem, īnsecūtus est mē cum suīs. clāmor, lapidēs, fustēs,
   gladiī. discessī in vestibulum Tettī Dāmōnis. quī erant mēcum
   facile operās aditū prohibuērunt. ipse occīdī potuit, sed ego
   diaetā cūrāre incipiō, chīrurgiae taedet.

3   **in prīmīs** especially.   **operās** gangs.
5   **ex senātūs cōnsultō** by decree of the senate.
7   **ante diem tertium Nōnās Novembrēs** on 3 November.
8   **fabrī** workmen.   **āreā** site.   **porticus Catulī:** the portico of Catulus
        was a public monument which had already been demolished once by Clodius's gangs.
10   **Quīntī . . . domus:** Quintus's house was next-door to Cicero's.
11   **īnspectante urbe = īnspectantibus cīvibus.**
12   **magnā querēlā et gemitū** with loud complaints and groans.
13–14   **ruere . . . cōgitāre . . . ostendere:** these are historic infinitives. Translate
        'he is rushing' etc.   **vel ante** even before.
16   **fustēs** clubs.
17   **vestibulum Tettī Dāmōnis** the forecourt of Tettius Damo's house.   **quī erant = eī quī erant.**
18   **diaetā** by a diet, a course of diet.
19   **chīrurgiae taedet** I'm tired of surgery.

70

1 What had happened to Cicero's house while he was in exile?
2 What did Clodius's gangs do to the workmen who were on Cicero's site?
3 What did they do next?
4 What happened to the house of Cicero's brother?
5 How did the people react to these outrages?
6 What was Clodius planning now?
7 'servīs apertē spem lībertātis ostendere' (l.14): why should Clodius do this?
8 Translate lines 15 to the end.
9 coniectū, iussū (ll.10–11): from what verb is each of these nouns formed and what does each noun mean?
10 ll.16–17 clāmor . . . gladiī: this sentence has no verb. What is the effect of this?
11 Explain what Cicero means by the last sentence (sed ego . . . taedet).
12 Comment on the situation in Rome which this letter reveals. How likely do you think similar outrages would be in our country today?

Exercise 6.4

Translate into Latin
1 When Cicero was consul, Catilina tried to overthrow the republic.
2 Cicero learnt what he was doing and accused him in the senate.
3 Catilina fled from Rome to join his allies in Etruria.
4 Cicero arrested the conspirators who had remained in Rome.
5 Everyone praised him and said that he had saved the republic.

Exercise 6.5

Translate the following passage into Latin in the form of a Latin letter.
(Cicero is writing to Terentia on his return from exile.)
Dearest Terentia,
                    I reached Brundisium on 5 August,
where I was greeted by Tullia, who had come to meet
me on her actual birthday. The next day I received a
letter from my brother Quintus, in which he said that
the law about my return had been passed by the people.
I set out for Rome yesterday and everywhere I have
been received by deputations who have come to
congratulate me. I hope to see you soon.
    Your loving husband, Tullius.

5 August **Nōnae Sextīlēs**   birthday **diēs nātālis**
I pass (a law) **perferō**   deputations **lēgātī, -ōrum,** *m.pl.*
to congratulate **grātulārī** + dative

CHAPTER VII

Cilicia and civil war

quīnque annōs Clōdius cum operīs tumultuābātur; neque senātus neque Pompēius furōrī eius moderārī poterat, sed T. Annius Milō, inimīcus Clōdī, quī prīmās partēs ēgerat in Cicerōne ab exsiliō revocandō, alterās operās comparāvit quibus Clōdiō
5 resisteret. saepe in viīs urbis, saepe in campō Martiō vī et armīs cōnflīgēbant. Clōdius vērō aedīlis factus summā impudentiā Milōnem in iūs vocāvit atque eum dē vī apud populum postulāvit. Cicerō in epistolā ad Quīntum frātrem rem sīc dēscrībit:

10   ante diem VIII Id. Febr. Milō adfuit. dīxit Pompēius, sīve voluit; nam, ut surrēxit, operae Clōdiānae clāmōrem sustulērunt, idque eī perpetuā ōrātiōne contigit nōn modo ut acclāmātiōne sed ut convīciō et maledictīs impedīrētur. quī ut perōrāvit, surrēxit Clōdius. eī tantus clāmor ā nostrīs (placuerat
15 enim referre grātiam) ut neque mente nec linguā neque ōre cōnsisteret. ea rēs ācta est, cum hōrā sextā vix Pompēius perōrāvisset, usque ad hōram VIII, cum omnia maledicta, versūs dēnique obscēnissimī in Clōdium et Clōdiam dīcerentur.
    ille furēns et exsanguis interrogābat suōs quis esset quī
20 plēbem fame necāret. respondēbant operae: 'Pompēius.' hōrā ferē nōnā quasi signō datō Clōdiānī nostrōs cōnspūtāre coepērunt. exārsit dolor. urgēre illī, locō ut nōs movērent. factus est ā nostrīs impetus; fuga operārum; ēiectus dē rōstrīs Clōdius; ac nōs quoque tum fūgimus, nē quid in turbā. senātus vocātus in
25 Cūriam. Pompēius domum.

    tandem cum rēs in tantum perturbātiōnis adductae essent ut trēs annōs vix fierī possent ipsa comitia, operae Clōdī Milōnisque forte convēnērunt in viā Appiā haud procul ab urbe; īnsecūta est caedēs in quā interfectus est Clōdius. postrīdiē Clōdiānī ad
30 furōrem excitātī corpus eius in forum ēlātum concremāvērunt ita ut flammae ventō lātae ipsam Cūriam dēlēvērunt. Milō autem in iūs vocātus atque ā iūdicibus condemnātus in exsilium fūgit. intereā Pompēius cōnsul sōlus factus est ut rempūblicam restitueret.

1	**tumultuābātur** ran riot.
2	**moderārī** (+ dative) to restrain.
3	**prīmās partēs** a leading part.
4	**alterās operās** another gang.
7–8	**apud populum postulāvit** prosecuted him before the people. Clodius as aedile prosecuted Milo before an assembly of the people in the Forum instead of in the ordinary courts. **dē vī:** this serious charge referred especially to violence in a political context, riot.
10	**ante diem VIII Id. Febr.** on 6 February. **Milō adfuit** Milo was present = appeared in court. Pompey was speaking for Milo.
10–11	**sīve voluit** or rather wanted to.
12	**perpetuā ōrātiōne** throughout his whole speech.
14	**perōrāvit** finished his speech. **clāmor** supply a verb, e.g. **factus est.**
14–15	**placuerat . . . grātiam** for we had decided to return the compliment. **placuerat** it had seemed good to us to . . .
15–16	**neque mente . . . cōnsisteret** he remained firm neither in mind . . . = he could control neither his mind . . .'.
18	**Clōdiam:** she was Clodius's sister, whose lovers included Caelius (see page 114) and the poet Catullus.
19	**exsanguis** bloodless = pale with anger.
19–20	**quī plēbem fame necāret:** Pompey was in charge of the city's corn supply.
21	**quasi** as if. **Clōdiānī** Clodius's men. **cōnspūtāre** to spit in unison at . . .
22	**exārsit dolor** tempers flared up. **urgēre illī** they shoved. **urgēre** is historic infinitive.
24	**nē quid . . .:** supply **accideret.**
26	**tantum perturbātiōnis** such chaos.
33	**Pompēius sōlus cōnsul:** rioting had prevented the consular elections taking place and so there had been no consuls. Pompey's appointment as sole consul was unique.

The forum and the Curia

35 in hōc discrīmine rērum, quia Pompēius novam lēgem dē
prōvinciīs tulerat, Cicerō ā senātū iussus est ad Ciliciam proficīscī
ut prōvinciam prōcōnsul administrāret. Māiō ineunte cum frātre
Quīntō fīliōque Mārcō Rōmā invītus discessit atque itinere
labōriōsō cōnfectō ad extrēmam prōvinciam mēnse Sextīlī
40 pervēnit.

prīmum in oppidīs complūribus iūs erat dīcendum ut iniūriās
prōvinciālium corrigeret, deinde ad castra festīnandum ut
Parthīs obstāre parāret, quī fluvium Euphrātem trānsgressī
fīnibus Ciliciae imminēbant. ad Atticum sīc scrībit:

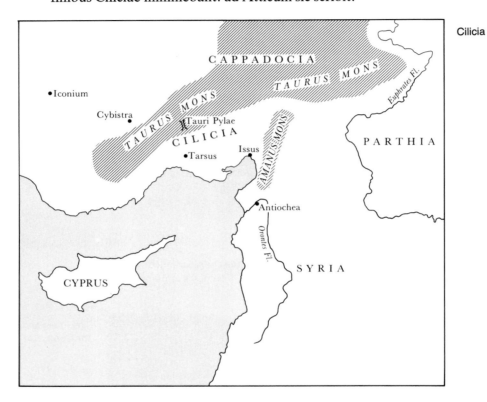

Cilicia

45 inde ad castra vēnī ante diem septimum Kalendās
Septembrēs. ante diem III exercitum lūstrāvī apud Iconium. ex
hīs castrīs, cum gravēs dē Parthīs nūntiī venīrent, perrēxī per
Cappadociae partem eam quae Ciliciam attingit. cum diēs
quīnque ad Cybistra Cappadociae castra habuissem, certior sum
50 factus Parthōs ab illō aditū Cappadociae longē abesse, Ciliciae
magis imminēre. itaque cōnfestim iter in Ciliciam fēcī per Taurī
pylās. Tarsum vēnī a.d. III Nōn. Octōbrēs. inde ad Amānum
contendī, quī Syriam ā Ciliciā in aquārum dīvortiō dīvidit; quī
mōns erat hostium plēnus sempiternōrum. hīc a.d. III Id. Oct.
55 magnum numerum hostium occīdimus. castella mūnītissima
cēpimus, incendimus. imperātōrēs appellātī sumus.

74

35	**in hōc discrīmine rērum** in this crisis of events.
37	**Māiō ineunte** at the beginning of May.
39	**extrēmam prōvinciam** the edge of his province.
41–2	**iniūriās prōvinciālium** wrongs done to the people of the province. Cicero's predecessor had been the brother of Clodius; he was responsible for many injustices.
45	**inde** from there; i.e. from Laodicea where Cicero had been administering justice.
45–6	**ante diem septimum Kal. Sept.** on 24 August.
46	**lūstrāvī** I reviewed.
47	**dē Parthīs:** since the defeat of Crassus at Carrhae (53 B.C.) the Romans had been continually expecting a Parthian invasion. **perrēxī** I proceeded, marched. Cappadocia was an independent kingdom on the northern border of Cilicia.
51–2	**Taurī pylās** the gates of Taurus: this was the only pass through the Taurus mountains. **Tarsum:** the birth place of St. Paul **a.d. III Nōn. Octōbrēs** on 5 October. **Amānum:** a mountain range (see map).
53	**in aquārum dīvortiō** on the watershed.
54	**sempiternōrum** perpetual. **a.d. III Id. Oct.** on 13 October.
56	**imperātōrēs appellātī sumus** I was hailed imperator. This title was only given to a general after a substantial victory when he was hailed '**imperātor**' by his troops. From now on he heads his letters **Cicerō imp**. It was up to the senate to decide whether such a victory merited a triumph in Rome, the highest honour a Roman could gain. Cicero was very keen to be granted a triumph but the civil war started before the Senate had made up its mind.

per Taurī pylās

75

castra paucōs diēs habuimus ea ipsa quae contrā Dārīum
habuerat apud Issum Alexander, imperātor haud paulō melior

Alexander equitēs in Dārīum dūcit apud Issum

quam aut tū aut ego. ibi diēs quīnque morātī, dīreptō et vastātō
60 Amānō inde discessimus. interim rūmōre adventūs nostrī et
Cassiō, quī Antiochīā tenēbātur, animus accessit et Parthīs timor
iniectus est. itaque eōs dēcēdentēs ab oppidō Cassius īnsecūtus
rem bene gessit. erat in Syriā nostrum nōmen in grātiā.

Cicerō cum Rōmā discessūrus esset, amīcum quendam,
65 M.Caelium Rūfum nōmine, rogāverat ut omnēs rēs urbānās
dīligentissimē sibi scrīberet. Caelius, iuvenis alacer et facētus,
omnia dēscrībit quae Rōmae geruntur et in rēbus sēriīs semper
aliquid iocī invenit. mēnse Septembrī hanc epistolam Cicerōnī
scrīpsit:

70 Caelius Cicerōnī s.d.
tantī nōn fuit Arsacēn capere et Seleucēam expugnāre, ut
eārum rērum quae hīc gestae sunt spectāculō carērēs; numquam
tibi oculī doluissent, sī in repulsā Domitī vultum vīdissēs. . .
dē summā rēpūblicā saepe tibi scrīpsī mē in annum pācem
75 nōn vidēre et, quō propius ea contentiō accēdit, eō clārius id
perīculum appāret. prōpositum hoc est dē quō quī rērum
potiuntur sunt dīmicātūrī, quod Cn.Pompēius cōnstituit nōn patī
C.Caesarem cōnsulem aliter fierī nisi exercitum et prōvinciās
trādiderit, Caesarī autem persuāsum est sē salvum esse nōn
80 posse, sī ab exercitū recesserit.
ad summam, quaeris quid putem futūrum esse. sī alter uter
eōrum ad Parthicum bellum nōn eat, videō magnās impendēre
discordiās, quās ferrum et vīs iūdicābit; uterque et animō et
cōpiīs est parātus. sī sine perīculō fierī posset, magnum et
85 iūcundum tibi Fortūna spectāculum parat.

76

<table>
<tbody>
<tr><td>58</td><td>**Issum:** see map. At Issus Alexander the Great had defeated Darius III of Persia in one of the decisive battles of world history, which left the Persian Empire open to him (333 B.C.).</td></tr>
<tr><td>61</td><td>**Cassiō . . . animus accessit** spirit was added to Cassius = Cassius plucked up heart. Cassius, who later with Brutus led the conspiracy against Julius Caesar, was at this time governor of Syria (capital Antioch).</td></tr>
<tr><td>63</td><td>**rem bene gessit** managed the affair well = won a success. He defeated the Parthian forces and killed their leader.   **erat . . . nostrum nōmen in grātiā** my name was in favour . . . = I was popular.</td></tr>
<tr><td>66</td><td>**alacer** lively.   **facētus** witty.</td></tr>
<tr><td>67</td><td>**in rēbus sēriīs** in serious matters.</td></tr>
<tr><td>68</td><td>**aliquid iocī** something funny.</td></tr>
<tr><td>71</td><td>**tantī nōn fuit** it was not worth so much.   **Arsacēn:** the king of Persia. **Seleucēam:** Seleucea was an important city on the Tigris in the heart of the Parthian empire.</td></tr>
<tr><td>73</td><td>**tibi oculī** the eyes for you = your eyes.   **in repulsā** at his rejection. Domitius was an exceptionally arrogant noble; he was unexpectedly beaten in an election.</td></tr>
<tr><td>74</td><td>**dē summā rēpūblicā** on the general political situation.</td></tr>
<tr><td>75</td><td>**quō propius . . . eō clārius** the nearer . . . the clearer . . .</td></tr>
<tr><td>76</td><td>**prōpositum hoc est** the issue is this . . .</td></tr>
<tr><td>76–7</td><td>**quī rērum potiuntur** those who control things, i.e. Pompey and Caesar.   **quī = eī quī.**</td></tr>
<tr><td>79</td><td>**Caesarī persuāsum est** it is persuaded to Caesar = Caesar is convinced (impersonal use of passive).</td></tr>
<tr><td>81</td><td>**ad summam** to sum up.   **alter uter** one or the other.</td></tr>
<tr><td>82</td><td>**impendēre** are threatening (us),</td></tr>
<tr><td>84</td><td>**sī fierī posset** if only it could happen . . .</td></tr>
</tbody>
</table>

Cicerō exercitum lūstrat

cum Cicerō prōvinciā mox discessūrus esset, litterās tamen
Caelī adhūc dēsīderābat; ad eum proximō annō sīc scrībit:

M.Cicerō Imp. s.d. M.Caeliō aedīlī curūlī.

sollicitus equidem eram dē rēbus urbānīs. ita tumultuōsae
90 cōntiōnēs, ita molestae Quīnquātrūs adferēbantur. illud molestē
ferō, nihil mē adhūc dē hīs rēbus habēre tuārum litterārum.
quārē etsī, cum haec legēs, ego iam annuum mūnus cōnfēcerō,
tamen obviae mihi velim sint tuae litterae quae mē ērudiant dē
omnī rēpūblicā, nē hospes plānē veniam. hoc melius quam tū
95 nēmō facere potest.

urbem, urbem, mī Rūfe, cole et in istā lūce vīve! omnis
peregrīnātiō, quod ego ab adulēscentiā iūdicāvī, obscūra et
sordida est iīs quōrum industria Rōmae potest illūstris esse. sed,
ut spērō, propediem tē vidēbō. tū mihi obviam mitte epistulās tē
100 dignās.

Cicerō ad Italiam regressus invēnit id quod Caelius
scrīpserat vērum esse; sīc scrībit ad Tīrōnem, scrībam suum,
quem in Graeciā relīquerat: 'ego ad urbem accessī prīdiē Nōnās
Iānuāriās. obviam mihi sīc est prōditum ut nihil posset fierī
105 ōrnātius. sed incidī in ipsam flammam cīvīlis discordiae vel potius
bellī.'

cum Caesar Rubicōnem trānsgressus Italiam invāsisset

Caesaris exercitus
Rubicōnem trānsit

Pompēiusque cum cōpiīs Brundisium contenderet ut in
Graeciam trānsīret, Cicerō ad summam dēspērātiōnem adductus
110 est. Caesarem reprehendit quod senātuī parēre nōluisset
exercitumque in Italiam dūxisset, Pompēium autem increpuit
quod Rōmā relictā bellum in aliās terrās extendere parāret. ad
Atticum sīc scrībit: 'uterque rēgnāre vult.' in vīllā Formiānā
cunctābātur, incertus quid sibi faciendum esset.

88	**aedīlī curūlī:** curule aedile was the full title of an aedile. Cicero heads this letter rather pompously, including his own title **imp.**
89	**sollicitus equidem eram** I was anxious (when I wrote this letter). We say 'I am anxious'.
89–90	**tumultuōsae cōntiōnēs** stormy meetings.
90	**molestae Quīnquātrūs** such a troublesome festival of Minerva. The Quinquatrus was a festival in honour of Minerva lasting from 19 to 23 March; it was a public holiday, in which violent political meetings had been held. **adferēbantur** are reported (to me).
90–1	**illud molestē ferō** I am upset at this.
91	**nihil tuārum litterārum** no letter from you.
92	**annuum mūnus** my year's duty. Cicero's term as governor was for a year and he left as early as he could.
93	**obviae mihi velim sint . . .** I should like a letter from you to meet me. **quae mē ērudiant** to inform me.
94	**plānē hospes** completely a stranger = a complete stranger. **veniam** I come (to Rome).
96	**in istā lūce:** the bright light of political life in Rome is contasted with the obscurity of foreign service.
97	**peregrīnātiō** foreign service. **quod iūdicāvī** as I have judged.
98	**illūstris esse** to be bright, shine out.
99	**propediem** soon.
102	**Tīrōnem:** Tiro was Cicero's confidential secretary; he had fallen ill on the return journey from Cilicia.
103–4	**prīdiē Nōnās Iānuāriās** on 4 January.
104	**obviam mihi est prōditum** it was come out to meet me = people came out . . . (impersonal use of passive).
104–5	**nihil posset fierī ōrnātius** nothing could have been done more honourable = I could not have had a more honourable reception.
105	**vel potius** or rather.
107	**Rubicōnem:** Caesar led his troops across the Rubicon, the border between his province and Italy on 10 January. Pompey immediately evacuated Rome and led his troops south to Brindisi.
111	**increpuit** blamed, criticized.
113	**in vīllā Formiānā:** Cicero had a villa at Formiae, on the bay of Naples.

V

appāreō (2)	I appear, am seen	**fluvius, -ī,** *m.*	river
certus-a-um	certain, sure, reliable	**haud**	not
certiōrem tē faciō	I inform you	**hospes, hospitis,** *c.*	guest, host; stranger
certior fīō	I am informed		
condemnō (1)	I condemn	**immineō** (2)	I hang over, threaten
contendō, -ere, contendī, contentum	I strain, strive, struggle, march, fight	+ dative	
		iniūria, -ae, *f.*	injury, wrong
		iūcundus-a-um	pleasant
contentiō, contentiōnis, *f.*	struggle, fight	**interrogō** (1)	I ask, question
contingit, -ere, contigit	it happens		
convīcium, -ī, *n.*	insult	**iūdicō** (1)	I judge
corrigō, -ere, corrēxī, corrēctum	I put straight, correct	**necō** (1)	I kill

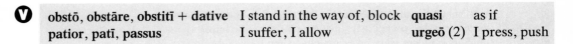

V | obstō, obstāre, obstitī + dative | I stand in the way of, block | **quasi** | as if |
| **patior, patī, passus** | I suffer, I allow | **urgeō** (2) | I press, push |

G Word building

Make sure that you know the following compounds of
iaciō, iacere, iēcī, iactum

abiciō, abicere, abiēcī, abiectum	I throw away
adiciō	I throw to, add to
dēiciō	I throw down, cast down
disiciō	I scatter
ēiciō	I throw out
iniciō	I throw in
obiciō	I throw in the way of
prōiciō	I throw forward, hurl
reiciō	I throw back, reject
subiciō	I throw under, place under
trāiciō	I convey across; I cross

Give English derivatives from as many of these compounds as you can.

G Conditional clauses

1 Open conditions

You have often met clauses of the form:

sī hoc fēcistī, prūdēns fuistī.
sī hoc facis, prūdēns es.
sī hoc fēceris, prūdēns eris.

Translate these examples

Conditional clauses of this sort, in which both Latin and English use
the indicative, are called open conditions; they leave open the
question of whether you did or did not do this.

2 Unfulfilled conditions

sī hoc fēcissēs, prūdēns fuissēs
If you had done this, you would have been sensible.

This is called an unfulfilled conditional clause, since it implies that you
did not do this. Latin uses the subjunctive; the pluperfect subjunctive
is used to refer to past time. Note that English uses a special
conditional tense indicated by 'should' or 'would'.

sī hoc facerēs, prūdēns essēs
If you were doing this, you would be sensible.

This is another unfulfilled conditional clause, since it implies that you are not doing this.

In unfulfilled conditional clauses referring to present time, Latin uses the imperfect subjunctive.

sī hoc faciās, stultus sīs

If you were to do this,
If you did this, you would be foolish.

This is called a remote future conditional clause; it refers to future time but suggests that you are unlikely to do this. Latin uses the present subjunctive in such clauses.

Exercise 7.1

Translate the following and say what type of conditional clause each is, i.e. past unfulfilled, present unfulfilled or remote future

1 sī statim vēnissēs, omnia tibi dīxissem.
2 sī frāter tuus adesset, nōs adiuvāret.
3 sī domum redeās, patrem videās.
4 sī pater vīveret, cōnsilium nōbīs daret.
5 nisi festīnāvissēmus, sērō advēnissēmus.
6 sī statim proficīscāmur, in tempore adveniāmus.

Exercise 7.2

In the following sentences both open and unfulfilled (and remote) conditional clauses occur. Before translating look carefully to see whether Latin uses the indicative or subjunctive

1 sī puerī bonī fuissent, magister fābulam nārrāvisset.
2 magister 'puerī,' inquit, 'sī dīligenter labōrētis, fābulam vōbīs nārrem.'
3 māter 'fīlia,' inquit, 'sī mē adiūveris, praemium tibi dabō.'
4 fīlia respondit: 'māter, sī nōn occupāta essem, tē libenter adiuvārem.'
5 nisi Pompēius novam lēgem dē prōvinciīs tulisset, Cicerō ad Ciliciam nōn iisset.
6 Cicerō cum Rōmā discēderet, Caeliō, 'sī quid novī audīveris,' inquit, 'epistulam mihi mitte.'
7 Caelius Cicerōnī scrībit: 'numquam tibi oculī doluissent, sī Domitī vultum vīdissēs.'
8 nōn patiēmur Caesarem cōnsulem fierī, nisi exercitum et prōvinciās trādiderit.
9 Caesar autem 'nōn salvus sim,' inquit, 'sī ab exercitū recēdam.'

Exercise 7.3

Translate lines 1–16 and answer the questions on the rest
Caesar, dum Pompēium Brundisium persequitur, hās litterās
Cicerōnī scrīpsit:

Caesar Imp. s.d. Cicerōnī Imp.

    cum Furnium nostrum tantum vīdissem neque loquī neque
5   audīre meō commodō potuissem, cum properārem atque essem
in itinere praemissīs iam legiōnibus, praeterīre tamen nōn potuī
quīn et scrīberem ad tē et illum mitterem grātiāsque agerem; ita
dē mē merēris. in prīmīs ab tē petō, quoniam cōnfīdō mē
celeriter ad urbem ventūrum, ut tē ibi videam, ut tuō cōnsiliō,
10 grātiā, dignitāte, ope omnium rērum ūtī possim; festīnātiōnī
meae brevitātīque litterārum ignōscēs. reliqua ex Furniō
cognōscēs.

    tandem tamen Cicerō cōnstituit ut sibi partēs Pompēiī
senātūsque nōn essent dēserendae. invītus Terentiam
15 Tulliamque valēre iussit, ad Graeciam profectūrus. simul atque
nāvem cōnscendit, hanc epistulam ad Terentiam mīsit:

    Tullius Terentiae suae s.p.

    nāvem spērō nōs valdē bonam habēre. in eam simul atque
cōnscendī, haec scrīpsī. deinde cōnscrībam ad nostrōs familiārēs
20 multās epistulās, quibus et tē et Tulliolam dīligentissimē
commendābō. cohortārer vōs ut animō fortiōrēs essētis, nisi vōs
fortiōrēs esse cognōssem quam quemquam virum. tū prīmum
valētūdinem tuam velim cūrāre; deinde, sī tibi vidēbitur, vīllīs iīs
ūtere quae longissimē aberunt mīlitibus. fundō Arpīnātī bene
25 poteris ūtī cum familiā urbānā, sī annōna cārior fuerit.

    Cicerō bellissimus tibi salūtem plūrimam dīcit. etiam atque
etiam valē.

    data VII Id. Iun.

3	**s.d. = salūtem dat (s.p. = salūtem plūrimam dat** l.17).
4	**Furnium:** Furnius was a tribune who was a friend of Cicero. **tantum** only.
5	**meō commodō** at my convenience, at my leisure. **cum** although.
6–7	**praeterīre nōn potuī quīn** I could not omit to . . .
8	**in prīmīs** in particular.
10	**grātiā** influence. **dignitāte** importance. **ope omnium rērum** help in everything.
13	**partēs** party, i.e. political party.
18	**nōs:** he had the young Marcus, his son, with him.
22	**cognōssem = cognōvissem.**
24	**ūtere:** imperative of **ūtor.** **fundō Arpīnātī** our farm at Arpinum.
25	**cum familiā urbānā** with the household from the city (Rome). **familia** includes all the slaves and servants. **annōna** the price of food.
28	**data VII Id. Iun.** posted 7 June.

1 What does Cicero hope?
2 What is he going to do when he has finished his letter to Terentia?
3 Translate: **cohortārer . . . virum** (lines 21–2). Why is the subjunctive used in these clauses?
4 Translate: **fundō Arpīnātī . . . fuerit** (lines 24–5). What tense is **fuerit**?
5 What does Cicero tell Terentia to do (two things)?
6 Why does he advise her to use the farm at Arpinum?
7 What do you think is the tone of this letter? How was Cicero feeling when he wrote it?

Exercise 7.4

Translate into Latin
1 When Cicero was fifty-five years old, he was sent by the senate to govern Cilicia.
2 He governed the province so well that he was praised by all the people.
3 Fearing that the Parthians would invade the province, he led his army to the borders.
4 After defeating the barbarians who lived in the mountains, he was called 'imperator' by his soldiers.
5 He hoped to earn a triumph, but did not know whether the senate would give him this honour.

earn **mereō** (2)

Exercise 7.5

Translate into Latin
When Cicero reached Cilicia, he found that the whole province had been oppressed by the man whom he had succeeded (**succēdō, -ere, -cessī, -cessum** + dative). And so he first had to administer justice in many cities in order to correct wrongs. He would have stayed longer in these cities, unless he had heard that the Parthians were threatening the borders of the province. And so he hurried to camp and prepared to march against them. After leading his army to the borders, he defeated the barbarians who lived in the mountains. The Parthians, fearing that Cicero was leading a great army against them, fled across the Euphrates.

Cicero – the last years

Cicerō cum ad Pompēī castra vēnisset, ad summam
dēspērātiōnem omnium rērum adductus est. Pompēius enim
īnfirmum sē praebēbat, nōbilēs autem arrogantēs ac crūdēlēs.
proeliō apud Pharsālum nōn interfuit, quia aeger erat. Pompēiō
5 victō ad Italiam regressus tōtum ferē annum Brundisī latēbat
miseriīs cōnfectus. tandem Caesarī ab Aegyptō redeuntī Tarentī
occurrit; ille, quī Cicerōnem et dīligēbat et colēbat, veniam
statim dedit. paulō post Cicerōnem, quī in vīllā Puteolīs
manēbat, rogāvit ut sē ad cēnam invītāret. Cicerō Atticō rem sīc
10 dēscrībit:

Cicerō Atticō sal.
     ō hospitem mihi gravem sed nōn molestum! fuit enim
periūcundē. sed cum secundīs Sāturnālibus ad Philippum vesperī
vēnisset, vīlla ita complēta est mīlitibus ut vix trīclīnium ubi
15 cēnātūrus ipse Caesar esset vacāret. sānē commōtus sum quid
futūrum esset postrīdiē; ac mihi Barba Cassius subvēnit,
custōdēs dedit. castra in agrō, vīlla dēfēnsa est.
     ille tertiīs Sāturnālibus apud Philippum ad hōram VII, nec
quemquam admīsit; ratiōnēs, opīnor, cum Balbō. inde ambulāvit
20 in lītore. post hōram VIII in balneum. unctus est, accubuit. et
ēdit et bibit iūcundē.
     praetereā tribus trīclīniīs acceptī comitēs valdē cōpiōsē.
lībertīs minus lautīs servīsque nihil dēfuit. nam lautiōrēs
ēleganter accēpī. quid multa? hominēs vīsī sumus. hospes tamen
25 nōn is cui dīcerēs, 'amābō tē, eōdem ad mē cum revertēris.'
semel satis est.
     Cicerō Rōmam regressus forō caruit. litterīs sē dēdidit
complūrēsque librōs dē arte rhētoricā et dē philosophiā
cōnscrīpsit. mala autem domestica eum valdē opprimēbant. cum
30 Terentiā enim discordia orta est, ex quā ēvēnit dīvortium. hanc
epistulam ad eam ultimam mīsit:

Tullius s.d. Terentiae suae
in Tusculānum nōs ventūrōs putāmus aut Nōnīs aut postrīdiē. ibi
ut sint omnia parāta. plūrēs enim fortasse nōbīscum erunt et, ut
35 arbitror, diūtius ibi commorābimur. labrum sī in balineō nōn est,
ut sit; item cētera quae sunt ad vīctum et ad valētūdinem
necessāria. valē.
     Kalendīs Octōbribus dē Venusīnō.

     paulō post mortua est Tullia, dēliciae eius. Cicerō maerōre

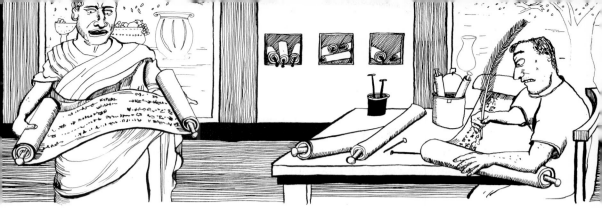

Cicerō librum Tīrōnī dictat

4	**proeliō apud Pharsālum:** the battle of Pharsalus took place in August 48 B.C. Cicero reached Brundisium in October and stayed there until the following September, not knowing how Caesar would treat him. After Pharsalus Caesar pursued Pompey to Egypt and was held up there until the following August.
12	**ō hospitem . . . nōn molestum** a formidable guest but not irksome to me.
12–13	**fuit periūcundē** he was in high spirits.
13	**secundīs Sāturnālibus:** on 18 December. The Saturnalia lasted from 17 to 25 December. Philippus, the stepfather of Octavian, had a villa near Cicero's.
14–15	**vix trīclīnium . . . vacāret** the dining room . . . could scarcely be kept clear.
15	**sānē** extremely.
16	**Barba Cassius:** one of Caesar's staff.
17	**castra in agrō:** supply **posita sunt**; this left Cicero's villa free of soldiers. Caesar had brought 2,000 with him.
18	**ad hōram VII:** supply **mānsit.**
19	**ratiōnēs (cōnficiēbat)** he was doing accounts. Balbus was his secretary and treasurer.
20	**unctus est** he was anointed. After a bath Romans anointed themselves with olive oil, which, in the dry Italian climate, kept their skin in good condition. **accubuit** reclined = took his place at table.
22	**acceptī:** supply **sunt.**
23	**minus lautīs** less grand.
24	**quid multa?** Why (should I say) much?; i.e. 'in short'. **hominēs vīsī sumus** I appeared (= showed myself) a man of the world.
24–5	**hospes tamen . . . dīcerēs** but he is not the sort of (**is**) guest to whom you would say . . .
25	**amābō tē, eōdem ad mē** please, (come) again (**eōdem** the same way) to me . . .
27	**forō caruit** avoided public life.
28	**complūrēs librōs:** between 46 and 44 B.C. Cicero wrote two works on oratory and eight on philosophy.
29	**mala domestica** family troubles.
30	**dīvortium** divorce.
33	**Tusculānum:** his villa at Tusculum, near Rome.
33–4	**ibi ut sint:** supply **cūrā** = see that.
35	**labrum** a basin. **in balineō** in the bathroom.
36	**item** in the same way: supply **cūrā ut sint. vīctum** sustenance, food.

40 cōnfectus est neque ūllā cōnsōlātiōne recreārī poterat. sīc ad
Atticum scrībit, dum in vīllā dēsertā prope mare manet:

Cicerō Atticō sal.
in hāc sōlitūdine careō omnium colloquiō, cumque māne mē in
silvam abstrūsī dēnsam et asperam, nōn exeō inde ante
45 vesperem. secundum tē nihil est mihi amīcius sōlitūdine. in eā
mihi omnis sermō est cum litterīs. eum tamen interpellat flētus,
cui repugnō quoad possum, sed adhūc parēs nōn sumus. Brūto,
ut suādēs, rescrībam. eās litterās crās habēbis.
        vetus amīcus eius, Servius Sulpicius, quī ex Asiā Athēnās
50 pervēnerat, sīc scrībit ut eum cōnsōlētur:
        posteā quam mihi nūntiātum est dē obitū Tulliae, fīliae
tuae, graviter molestēque tulī, commūnemque eam calamitātem
exīstimāvī; sī istīc adfuissem, neque tibi dēfuissem cōramque
meum dolōrem tibi dēclārāssem.
55   ex Asiā rediēns cum ab Aegīnā Megaram versus nāvigārem,
coepī regiōnēs circumcircā prōspicere. post mē erat Aegīna, ante
mē Megara, dextrā Pīraeus, sinistrā Corinthus, quae oppida
quōdam tempore flōrentissima fuērunt, nunc prōstrāta et dīruta
ante oculōs iacent. coepī mēcum sīc cōgitāre. 'hem! nōs
60 homunculī indignāmur, sī quis nostrum interiit aut occīsus est,
quōrum vīta brevior esse dēbet, cum ūnō locō tot oppidōrum
cadāvera prōiecta iacent? vīsne tū tē, Servī, cohibēre et
meminisse hominem tē esse nātum?'

        praetereā dē fīliō Mārcō valdē ānxius erat, quī Athēnās
65 missus ut in Lyceō apud Cratippum studēret, summā licentiā
ūtēbātur maximōsque faciēbat sūmptūs. amīcus Cicerōnis, C.
Trebōnius, Mārcō vīsō dum ad Asiam iter facit, hanc epistulam
ad Cicerōnem mīsit, quae aliquid speī eī attulit:
C. Trebōnius Cicerōnī s.d.
70 sī valēs, bene est. Athēnās vēnī ante diem XI Kal. Iūn. atque ibi,
quod maximē optābam, vīdī fīlium tuum dēditum optimīs studiīs
summāque modestiae fāmā. quā ex rē quantam voluptātem
cēperim, scīre potes mē tacente. nōlī putāre, mī Cicerō, mē hoc
auribus tuīs dare; nihil adulēscente tuō aut amābilius omnibus iīs
75 quī Athēnīs sunt aut studiōsius eārum artium quās tū maximē
amās.
        nihilōminus Cicerō dēcrēvit Athēnās iter facere ut Mārcum
ipse vīseret. mēnse Quīntīlī profectus, in Siciliam pervēnerat
cum nūntium accēpit quī eī persuāsit ut Rōmam redīret.
80   iamdūdum forō caruerat neque ūllam partem in
coniūrātiōne Brūtī ēgerat. nunc tamen veritus nē Antōnius
rēpūblicā ēversā rēgnum occupāret, Rōmam invītus rediit ut
lībertātem dēfenderet.
        in Antōnium et in senātū et apud populum identidem
85 vehementer invectus est; omnēs omnis ordinis quī reīpūblicae

43	**careō omnium colloquiō** I lack conversation with all = I don't talk to anyone. **māne** early in the morning.
43–4	**mē abstrūsī** I have hidden myself.
45	**secundum tē** second to you, after you. **amīcius** more dear.
46	**litterīs** literature, books. **interpellat** interrupts.
47	**quoad** as far as. **Brūtō:** Brutus had written a letter of condolence. Cicero intends to answer it and send a copy of his answer to Atticus.
52	**graviter molestēque tulī** I bore it seriously and painfully = I was extremely upset.
53	**cōram** in your presence, i.e. in person, not by letter.
55	**Megaram versus** towards Megara. All the places mentioned in this passage had been important in earlier times but had been in ruins since the Greeks rebelled against the Romans 100 years before.
58	**prōstrāta et dīruta** thrown down and demolished.
59–60	**hem! nōs homunculī indignāmur** come! are we poor men indignant?
61	**quōrum vīta . . . dēbet:** i.e. the life of a man is bound to be shorter than that of a city.
62	**cadāvera** corpses. **vīsne tē cohibēre** won't you take control of yourself?
63	**hominem** a human being.
65–6	**summā licentiā ūtēbātur** was using great licence = was living very wildly.
66	**maximōs sūmptūs faciēbat** he was making very great expenses = he was being very extravagant.
70	**sī valēs, bene est** if you are well, that's good. This is a polite formula for starting a letter; compare 'I hope you are well'. **ante diem XI Kal.Iun.** 22 May.
71	**quod** which thing = as.
72	**summā modestiae fāmā** with the highest reputation for good behaviour.
73–4	**mē hoc auribus tuīs dare** am giving this to your ears = saying this to please you. **nihil** no one. Supply **est.**
80	**iamdūdum forō:** the date is July 44 B.C. Caesar was murdered on 15 March; Octavian and Antony are struggling for power; Brutus and Cassius have just been driven from Italy. Cicero made his first attack on Antony in the senate on 2 September. From December he rallied the senate against Antony and made a series of speeches against him, of which twelve survive.
82	**rēgnum** absolute power.

Sulpicius Megaram versus nāvigat

lībertātīque studēbant cohortātus est ut Antōniō resisterent;
Brūtum et Cassium ad bellum incitāvit; Octāviānum in partēs
senātūs indūcere cōnābātur. sed frūstrā contendēbat. namque eō
ipsō tempore cum, Antōniō ad Mutinam victō, rēs prosperē
90   gerere vidēbātur, Octāviānus Cicerōne senātūque prōditīs
foedus cum Antōniō Lepidōque fēcit. triumvirī Rōmam
occupāre, omnia ad suam voluntātem agere, inimīcōs suōs
prōscrībere. Antōnius ipse mīlitēs mīsit quī Cicerōnem
interficerent.

(V)

calamitās, calamitātis, *f.*	disaster	**miseria, -ae,** *f.*	misery
cēdō, -ere, cessī, cessum	I yield, give way, depart	**nihilōminus**	nevertheless
		optō (1)	I wish for, pray for
cōnstat, cōnstāre, cōnstitit (impersonal)	it is agreed	**pār, paris**	equal
cōpia, -ae, *f.*	plenty; supply	**prō certō habeō**	I am certain
cōpiōsus-a-um	plentiful	**revertor, -ī, reversus**	I return
dēclārō (1)	I make clear, declare	**sōlus-a-um**	alone
		sōlitūdō, sōlitūdinis, *f.*	solitude, loneliness
hortor (1)	I encourage		
cohortor (1)	I encourage	**subveniō, -īre, subvēnī, subventum** + dative	I come to help
intueor (2)	I gaze at		
iussū	by order	**ultimus-a-um**	last; furthest
lateō (2)	I lie hidden		

(G) Word building

Make sure that you know the following compounds of **speciō, specere, spexī**

aspiciō, aspicere, aspexī, aspectum	I look at	**dēspiciō**	I look down on, despise
		prōspiciō	I look forward, look out for
cōnspiciō	I catch sight of	**respiciō**	I look back (at), consider

Give English derivatives from as many of these compounds as you can.

(G) Alternative spellings

Note the following:
1 2nd declension masculine nouns with nominative ending **-ius** and
   neuter nouns with nominative ending **-ium**, e.g. **fīlius, ingenium**:
   genitive singular: **fīliī** or **fīlī, ingeniī** or **ingenī**.
2 From **is, ea, id**: nominative masculine plural: **eī** or **iī**; dative and
   ablative plural **eīs** or **iīs**.
3 **epistola** or **epistula**; **cotīdiē** or **cottīdiē**; **numquam** or **nunquam**.
4 shortened forms: **mī** for **mihi**; **amāram** for **amāveram, amāsse** for
   amāvisse, amāssem for **amāvissem** etc.

92–3   **occupāre . . . agere . . . prōscrībere:** historic infinitives. Translate as
        indicatives: the triumvirs occupied etc.
93   **prōscrībere** outlawed.

Lepidus   Octāviānus   Antōnius

Exercise 8.1

Translate

1 frātris valētūdō nōbīs magnae cūrae est. quid nōbīs faciendum est?
2 vōs oportet servum mittere quī patrem dē rē certiōrem faciat.
3 sī hoc faciāmus, pater valdē īrāscātur.
4 nōnne nōbīs ipsīs domum redeundum est?
5 sī herī profectī essēmus, iam domum pervēnissēmus.
6 nōlīte morārī; licet vōbīs hodiē proficīscī.
7 vīsne tū nōbīscum venīre? tū, vir summā prūdentiā, magnō auxiliō
   nōbīs eris.
8 veniam vōbīscum, sī equum mihi dederitis.
9 cum hoc placuisset, omnia parāvērunt ad iter faciendum.
10 cum tertiā hōrā urbe discessissent, celeriter domum perventum
   est.

Exercise 8.2

Translate

1 ad nōbilium voluntātem mihi conciliandam maximō tē mihi ūsuī
   fore videō.
2 valdē fuit Quīntō properandum, nē quid absēns acciperet
   calamitātis.
3 cohortārer vōs ut animō fortiōrēs essētis, nisi vōs fortiōrēs esse
   cognōvissem quam quemquam virum.
4 pater noster, cum īnfirmā esset valētūdine, hīc aetātem ēgit in
   litterīs.
5 vōs videō esse miserrimās, quae, nisi ego timidus fuissem, beātae
   fuissētis.

89

Exercise 8.3

Answer the questions below this passage without translating except where you are asked to

Cicerōnis mortem sīc dēscrībit T. Līvius:

M. Cicerō sub adventum triumvirōrum cesserat urbe, prō certō habēns nōn magis Antōniō ēripī sē quam Caesarī Cassium et Brūtum posse. prīmō in Tusculānum fūgit, inde trānsversīs
5 itineribus in Formiānum, ut ab Cāiētā nāvem cōnscēnsūrus, proficīscitur. unde aliquotiēns in altum prōvectum cum modo ventī adversī retulissent, modo ipse iactātiōnem nāvis patī nōn posset, taedium tandem eum et fugae et vītae cēpit; regressusque ad villam, quae paulō plūs mīlle passibus ā marī abest, 'moriar'
10 inquit 'in patriā saepe servātā.'

satis cōnstat servōs fortiter fidēliterque parātōs fuisse ad dīmicandum; ipsum dēpōnī lectīcam et quiētōs patī quod fors inīqua cōgeret iussisse. prōminentī ex lectīcā praebentīque immōtam cervīcem caput praecīsum est. nec satis stolidae
15 crūdēlitātī mīlitum fuit. manūs quoque, scrīpsisse in Antōnium aliquid exprobrantēs, praecīderunt.

ita relātum caput ad Antōnium iussū eius inter duās manūs in rōstrīs positum, ubi ille cōnsul, ubi saepe cōnsulāris, ubi eō ipsō annō adversus Antōnium cum admīrātiōne ēloquentiae
20 audītus fuerat. vix attollentēs prae lacrimīs oculōs hominēs intuērī trucīdāta membra eius poterant.

1	**T. Livius:** the historian, from whose work you will later read extracts.
2	**sub** towards = just before.
3	**nōn magis Antōniō ēripī sē ...** that he could no more be saved from Antony ... **Caesarī** = Octavian. Cicero was Antony's chief enemy, Brutus and Cassius Octavian's.
4	**Tusculānum:** Cicero's Tusculan villa was near Rome. His villa at Formiae was on the Bay of Naples and Caieta was a little port west of Formiae. **trānsversīs** cross-country.
5	**ut ... cōnscēnsūrus** intending to board ...
6	**in altum prōvectum:** supply **eum**, the object of **retulissent.**
7	**iactātiōnem** tossing.
8	**taedium ... et fugae et vītae cēpit** weariness of both flight and life took him ... = he became tired of both fleeing and living.
11–12	**ad dīmicandum** to fight.
12–13	**ipsum dēpōnī lectīcam ... iussisse** = (sē) ipsum iussisse lectīcam dēpōnī et eōs ...; accusative and infinitive because it is part of what was agreed (**satis cōnstat**).
12–13	**fors inīqua** cruel fate.
13–14	**prōminentī ... cervīcem** to him sticking out and keeping immobile his neck = while he stuck out ...
14	**stolidae** brutal.
16	**exprobrantēs** saying insultingly that (these hands) ...
18	**cōnsulāris** as ex-consul. Consulars were the senior statesmen of Rome.
20	**prae lacrimīs** because of their tears.
21	**trucīdāta** butchered.

1 Why did Cicero leave Rome?
2 Where did he go?
3 What prevented him from escaping by sea?
4 What were his last recorded words?
5 How did the slaves react to the situation?
6 What did Cicero tell them to do?
7 How did the soldiers show 'stolida crūdēlitās'?
8 Translate the last paragraph.
9 How well, in your opinion, did Cicero meet his death?
10 Write a description of Cicero's character, taking into account all the letters of his which you have read.

The rōstra

Exercise 8.4

Translate into Latin

1 When Cicero returned from Cilicia to Rome, he was vexed by many troubles.
2 He divorced Terentia, with whom he had lived for very many years.
3 When his daughter Tullia died, he was so sad that he retired to a deserted villa.
4 When he heard that his son Marcus was behaving badly in Athens, he set out for Greece in order to see him.
5 Returning to Italy, he led the senate against Antonius and would have saved the republic, if Octavianus had not betrayed him.

divorce **repudiō** (1)   I behave **mē gerō**

Exercise 8.5

Translate into Latin, in the form of a Latin letter

Dear Marcus,

      I have received your letter, in which you say that you are again short of money. I told Tiro to send you 10,000 sesterces last month. I do not know how you run up such great expenses. How are you using your money? I hear that you are not working hard but waste a lot of time drinking in pubs. I am extremely worried about you and will soon set out for Greece to visit you myself.

   Your loving father
          Cicero.

I am short of **careō** (2) + ablative   waste **terō, -ere**
sesterce **sēstertius, -ī,** *m.*   pub **taberna, -ae,** *f.*
run up **faciō**

CATULLUS

CHAPTER IX

Catullus and his friends

C. Valerius Catullus Vērōnae nātus est cōnsulibus L. Cornēliō
Cinnā, Cn Octāviō, quod oppidum erat flōrentissimum in Galliā
Trānspadānā. pater eius vir locuplēs erat, amīcus Iūliī Caesaris,
quī apud eum manēbat cum Galliam Cisalpīnam vīsitāret ad iūs
5 dīcendum. pater Catullī vīllam pulchram possidēbat in
paenīnsulā Sirmiōne in lītore lacūs Bēnācī. illīc tōta familia
quotannīs per aestātem manēbat; Catullus hunc locum valdē
dīligēbat. cum illūc rediisset longam post peregrīnātiōnem,
vīllam sīc salūtāvit:

10   paene īnsulārum, Sirmiō, īnsulārumque
        ocelle, quāscumque in liquentibus stagnīs
        marīque vastō fert uterque Neptūnus,
        quam tē libenter quamque laetus invīsō,
        vix mī ipse crēdēns Thūniam atque Bīthūnōs
15   līquisse campōs et vidēre tē in tūtō.
        ō quid solūtīs est beātius cūrīs,
        cum mēns onus repōnit, ac peregrīnō
        labōre fessī vēnimus larem ad nostrum
        dēsīderātōque acquiēscimus lectō?
20   hoc est quod ūnum est prō labōribus tantīs.
        salvē, ō venusta Sirmiō, atque erō gaudē
        gaudente, vōsque, ō Lȳdiae lacūs undae,
        rīdēte quidquid est domī cachinnōrum.   (31)

1 Explain the meaning of the first three lines in simple terms.
2 Translate lines 13–15 (**quam tē . . . in tūtō**).
3 Summarize the meaning of lines 16–20 (**ō quid . . . tantīs**).
4 What does Catullus mean when he tells the waves to laugh (l.23)?
5 How far does this poem seem to you the natural expression of natural
  feelings? Have you ever had such feelings yourself? If so, when?
6 At what points does the poem seem to you to become artificial in expression?
  Why should Catullus have chosen to write like this at these points?

        dē pueritiā eius nihil cognōvimus. vīgintī ferē annōs nātus
25 Rōmam migrāvit et mox multōs habēbat amīcōs inter iuvenēs
lautōs quī litterīs dēditī erant. Catullus ipse versūs scrībere
incipiēbat. quōdam tempore cum Licinius Calvus, iuvenis
īnsignis, eum ad cēnam invītāvisset, per vīnum uterque versūs in
vicem scrībēbat. Catullus adeō sē oblectāverat ut posterō diē hoc
30 poēma ad Calvum mīserit:

1	**Vērōnae:** Verona, a prosperous town, was in Transpadane Gaul, i.e. Gaul the other side of the river Padus (modern Po). **L. Cornēliō . . . cōnsulibus:** i.e. in the year 87 B.C.
6	**paenīnsulā:** Sirmio is a small rocky peninsula on the south shore of Lake Garda, about twenty miles from Verona. **paenīnsula** an almost island, an idea Catullus plays with in line 1 of the poem.
11	**ocelle: ocellus** means 'a little eye', an endearing diminutive. 'Eye of all islands' means here 'jewel . . .' **liquentibus stagnīs** clear lakes.
12	**uterque Neptūnus** each Neptune. Catullus speaks as if there were two Neptunes, one presiding over sea water, the other over fresh water.
13	**quam libenter** how gladly.
14–15	**mī = mihi. Thūniam atque Bīthūnōs campōs:** Catullus has just returned from a year's service on the staff of the governor of Bithynia, where the people consisted of two tribes, the Thūnī and the Bīthūnī.
15	**līquisse** that I have left (**mē** is omitted).
16	**solūtīs cūrīs** than cares untied = than throwing off one's cares
17–8	**peregrīnō labōre** toils in foreign lands.
18	**larem** home.
20	**hoc est . . . tantīs** this is the one thing which makes up for such great toils.
21–2	**venusta** lovely. **erō gaudente** when your master rejoices.
22	**Lȳdiae undae:** he calls the waves of the lake Lydian because the district round Lake Benacus (Garda) was originally settled by Etruscans, who were thought to have come from Lydia (modern Turkey).
23	**rīdēte . . . cachinnōrum** laugh with all the noisy laughter (**cachinnōrum**) you have at your command (literally: whatever of noisy laughter there is in your home).
25	**migrāvit** moved.
26	**lautōs** smart. **dēditī** devoted.
28	**per vīnum** over the wine.
28–9	**in vicem** in turn.
29	**sē oblectāverat** had enjoyed himself.

vōsque, ō Lȳdiae lacūs undae, rīdēte   Lake Garda (Bēnācus)

hesternō, Licinī, diē ōtiōsī
multum lūsimus in meīs tabellīs,
ut convēnerat esse dēlicātōs;
scrībēns versiculōs uterque nostrum
35   lūdēbat numerō modo hōc modo illōc,
reddēns mūtua per iocum atque vīnum.
atque illinc abiī tuō lepōre
incēnsus, Licinī, facētiīsque,
ut nec mē miserum cibus iuvāret
40   nec somnus tegeret quiēte ocellōs,
sed tōtō indomitus furōre lectō
versārer, cupiēns vidēre lūcem,
ut tēcum loquerer simulque ut essem.
at dēfessa labōre membra postquam
45   sēmimortua lectulō iacēbant,
hoc, iūcunde, tibī poēma fēcī,
ex quō perspicerēs meum dolōrem.
nunc audāx cave sīs, precēsque nostrās,
ōrāmus, cave dēspuās, ocelle,
50   nē poenās Nemesis reposcat ā tē.
est vēmēns dea. laedere hanc cavētō.   (50)

puella tabellās tenet

1 Explain in your own words how Catullus and Calvus (= Licinius)
spent the evening?
2 How did Catullus feel when he got home?
3 **cupiēns vidēre lūcem** (l.42): translate and explain why Catullus
wanted this.
4 **dolōrem** (l.47): what pain is he talking about?
5 Translate the last four lines.
6 Show how the language of Catullus suggests that he has fallen in
love with Calvus. How serious do you suppose he is? (Calvus was
short and fat; if it wasn't his beauty that Catullus had fallen for,
what else might it have been?)
7 **precēs nostrās** (l.48): what do you suppose was Catullus's prayer?

    brevī tempore Calvus Catullī familiāris intimus factus est.
iuvenis erat alacer et facētus, corpore exiguō, ōrātor disertus.
cum Vatīnium accūsāret, Caesaris fautōrem sed plērīsque
55   invīsum, Calvus tantā ēloquentiā causam agēbat ut Vatīnius
dēspērāns iūdicibus 'rogō vōs, iūdicēs,' inquit, 'num sī iste
disertus est, ideō mē oportet damnārī?' Catullus, quī in corōnā
stābat Calvumque summā cum admīrātiōne audiēbat, hōs
versiculōs dē amīcō suō scrīpsit.

31   **hesternō diē** yesterday.

32   **tabellīs** tablets, i.e. notebook; **tabella** is a diminutive of **tabula**.
Catullus is very fond of diminutive forms; in this poem we have
versiculōs (l.34) little verses, **ocellōs** (l.40) little eyes, poor eyes, **lectulō**
(l.45) little bed, **ocelle** (l.49), which here has another meaning 'my
darling'. Diminutives can express smallness, affection, pity, contempt,
according to the context.

33   **ut convēnerat esse dēlicātōs** as it had been agreed (= we had agreed) to
be smart.

35   **numerō modo hōc modo illōc** now in this metre, now in that.

36   **reddēns mūtua** paying back things in turn, i.e. they chose a subject and
a metre and then wrote epigrams (short poems) in turn, each, so to
speak, paying back the other with a new poem.

37–8   **lepōre . . . facētiīsque** by your charm and wit. Catullus uses the
language of love: **incēnsus, miserum** etc. **miser** is often used of
unhappy lovers (they can neither sleep nor eat!).

41   **indomitus furōre** uncontrolled in my frenzy.

42   **versārer** tossed and turned.

43   **simulque ut essem** and to be with you.

48   **nunc audāx cave sīs** now beware of being rash.

49   **cave dēspuās** beware of spitting out, i.e. rejecting.

50   Nemesis was the goddess of vengeance, who punished those who were
proud.

51   **vēmēns = vehemēns. cavētō** a solemn form of the imperative for **cavē**
(compare **estō, scītō**).

multum lūsimus in meīs tabellīs

52   **intimus** closest.

53   **alacer** lively.   **corpore exiguō** of small, tiny stature.

54   **fautōrem** a supporter.

57   **ideō** for that reason.   **corōnā** ring. Roman trials were held in the open
air, in the Forum. They were public events which people attended in
crowds for entertainment; they stood round the court (presiding
magistrate, jury of seventy-five **iūdicēs**, three defence and three
prosecuting counsel, witnesses and friends of the accused). Calvus
prosecuted Vatinius twice, the second time in 54 B.C. when Cicero
successfully defended him.

60 rīsī nescioquem modo ē corōnā,
qui, cum mīrificē Vatīniāna
meus crīmina Calvus explicāsset,
admīrāns ait haec manūsque tollēns,
'dī magnī, salapūtium disertum!'   (53)

65   Sāturnālibus quondam Calvus mūnus novum ad Catullum
per iocum mīsit, libellum poētārum pessimōrum. Catullus,
simulāns sē valdē indignārī, hoc carmen rescrībit:

nī tē plūs oculīs meīs amārem,
iūcundissime Calve, mūnere istō
70 ōdissem tē odiō Vatīniānō:
nam quid fēcī ego quidve sum locūtus
cūr mē tot male perderēs poētīs?
istī dī mala multa dent clientī,
quī tantum tibi mīsit impiōrum.
75 quod sī, ut suspicor, hoc novum ac repertum
mūnus dat tibi Sulla litterātor,
nōn est mī male, sed bene et beātē,
quod nōn dispereunt tuī labōrēs.
dī magnī, horribilem et sacrum libellum!
80 quem tū scīlicet ad tuum Catullum
mīstī, continuō ut diē perīret,
Sāturnālibus, optimō diērum!
nōn nōn hoc tibi, salse, sīc abībit.
nam sī lūxerit, ad librāriōrum
85 curram scrīnia, Caesiōs, Aquīnōs,
Suffēnum, omnia colligam venēna,
ac tē hīs suppliciīs remūnerābor.
vōs hinc intereā valēte, abīte
illūc unde malum pedem attulistis,
90 saeclī incommoda, pessimī poētae.   (14)

Sulla litterātor

    Catullus iuvenis erat animō mōbilī atque ārdentī; omnēs
quōs cognōverat aut amābat aut ōderat. Iūlium Caesarem,
familiārem patris suī, nōn valdē dīligēbat; cum ille Catullum
rogāvisset ut sē in aliquā rē iuvāret, sīc respondit Catullus:

95 nīl nimium studeō, Caesar, tibi velle placēre,
nec scīre utrum sīs albus an āter homō.   (93)

    Cicerōnem autem admīrārī vidēbātur atque colere; cum
Cicerō eum in causā aliquā adiūvisset, ille grātiās sīc ēgit:

disertissime Rōmulī nepōtum,
100 quot sunt quotque fuēre, Mārce Tullī,
quotque post aliīs erunt in annīs,

60	**rīsī nescioquem** I laughed at some man. **modo** recently.
61–62	**mīrificē explicāsset** had explained marvellously.
64	**salapūtium disertum** what an eloquent little cock!
65	**Sāturnālibus quondam** on the Saturnalia once. At the feast of Saturn, which began on 17 December, friends sent each other presents. Our Christmas celebrations are descended from this festival which ended on 25 December.
68	**nī** = **nisi.**
69	**mūnere istō** because of that gift.
70	**odiō Vatīniānō** with the hatred Vatinius feels for you.
71	**quidve** or what?
73	**istī dī ... dent clientī** may the gods give ... to that client. Catullus guesses that the gift was a present to Calvus from a client.
74	**tantum impiōrum** such a pack of sinners.
75	**quod sī** but if. **repertum** recherché, rare.
76	**Sulla litterātor** Sulla the schoolmaster. Catullus guesses that the client was Sulla, whom Calvus had helped in some way.
77	**nōn est mī male ... beātē** it is not bad for me but well and happy = I'm not upset but very pleased.
78	**dispereunt** perish = are wasted.
79	**sacrum** cursed. **libellum** accusative of exclamation.
80	**scīlicet** I suppose.
81	**mīstī** = **mīsistī.**
83	**nōn ... abībit** it won't end. **salse** my witty friend.
84–5	**sī lūxerit** as soon as it is light. **librāriōrum scrīnia** the shelves (literally boxes) of the booksellers.
85	**Caesiōs, Aquīnōs** men like Caesius, Aquinus. These were bad poets, as was Suffenus.
87	**remūnerābor** I will pay you back.
88	**vōs**: to whom is he speaking?
89	**unde malum pedem attulistis** from where you brought your bad feet. This means two things at once: 'from where you limped your way here' and 'from where you brought your bad verse (metrical feet)'.
90	**saeclī incommoda** curse of our time.
91	**mōbilī** excitable. **ārdentī** burning, passionate.
95	**nīl nimium studeō** I am not really keen.
96	**albus an āter** white or black.
99	**Rōmulī nepōtum** the descendants of Romulus: a rather grand way of saying 'the Romans'.
100	**quot** as many as = all who. **fuēre** = **fuērunt. Mārce Tullī** = Marcus Tullius Cicero.

sī iste disertus
est, ideō mē
oportet damnārī?

<pre>
        grātiās tibi maximās Catullus
        agit pessimus omnium poēta,
        tantō pessimus omnium poēta,
105   quantō tū optimus omnium patrōnus.   (49)
</pre>

          cum amīcus quīdam, nōmine Verānius, ab Hispāniā Rōmam
rediisset, Catullus eum sīc salūtāvit.

<pre>
        Verānī, omnibus ē meīs amīcīs
        antistāns mihi mīlibus trecentīs,
110   vēnistīne domum ad tuōs penātēs
        frātrēsque ūnanimōs anumque mātrem?
        vēnistī. ō mihi nūntiī beātī!
        vīsam tē incolumem audiamque Hibērum
        nārrantem loca, facta, nātiōnēs,
115   ut mōs est tuus, applicānsque collum
        iūcundum ōs oculōsque suāviābor?
        ō quantum est hominum beātiōrum,
        quid mē laetius est beātiusve?   (9)
</pre>

V

beātus-a-um	blessed, happy	**disertus-a-um**	eloquent
candidus-a-um	shining, white, beautiful	**facētus-a-um**	witty
		lacus, -ūs, *m.*	lake
caveō, cavēre, cāvī, cautum	I beware	**lectus, -ī,** *m.*	couch, bed
		membrum, -ī, *n.*	limb
cibus, -ī, *m.*	food	**precēs, precum,** *f.pl.*	prayers
continuō (adverb)	straightway	**quisquis, quidquid**	whoever, whatever
damnō (1)	I condemn	**-ve**	or

G Word building

Make sure that you know the following compounds of **mittō, mittere, mīsī, missum**

admittō	I send to, admit; I commit	**immittō**	I send in
āmittō	I lose	**omittō**	I let go, give up
committō	I entrust	**permittō**	I allow
proelium committō	I join battle	**praemittō**	I send ahead
dēmittō	I send down, let down	**prōmittō**	I promise
dīmittō	I dismiss	**remittō**	I send back, relax
ēmittō	I send out		

Give English derivatives from as many of these compounds as you can.

104–5	**tantō pessimus . . . quantō optimus** so much the worst . . . as you are the best. **patrōnus** advocate.
106	**Verānius:** Veranius had been serving on the staff of the governor in Spain.
109	**antistāns . . . mīlibus trecentīs** of all my friends excelling 300,000 for me = worth more to me than 300,000 of my friends. 300 is used of an indefinitely large number; we say 'thousands', or even 'millions', in such contexts.
110	**penātēs** home.
111	**ūnanimōs** of one mind = loving. **anum** old.
112	**ō nūntiī beātī** o happy message (**nūntiī** is poetic plural for singular).
113	**Hibērum:** genitive plural, 'of the Spaniards'.
115	**applicāns collum** drawing your neck (to me) = putting my arm round your neck.
116	**suāviābor** I shall kiss.
117	**ō quantum . . .** oh, whatever there is of happier men = oh, of all the happy men there are.
118	**quid: quid** (what? is used to mean 'who?'. Hence **laetius** and **beātius** are neuter, although they refer to Catullus.

Diminutives

Diminutive forms primarily express smallness, but they can also convey the ideas of pity, affection, or contempt, according to the context.

Examples of diminutives

fīlia	**fīliola**	a little daughter	**homō**	**homunculus**	a little man
fīlius	**fīliolus**	a little son	**māter**	**mātercula**	a little mother
ager	**agellus**	a little field	**versus**	**versiculus**	a little verse
liber	**libellus**	a little book	**miser**	**misellus**	poor little
oculus	**ocellus**	a little eye	**parvus**	**parvulus**	dear little

The subjunctive in main clauses

The indicative mood is used to express facts (and questions), the imperative mood to express commands. The subjunctive mood, in main clauses, is used for various forms of non-factual expression.

Indicative: **puerī ad lūdum festīnant** The boys are hurrying to school.
  nōnne festīnant puerī? Aren't the boys hurrying?
Imperative **festīnāte, puerī** Hurry, boys.
Subjunctive
1 **festīnēmus, puerī** Let's hurry, boys.
  festīnent puerī Let the boys hurry.
  nē domī maneāmus Let us not stay at home.

First and third person commands or encouragement; negative **nē** (jussive subjunctive).

2 redeant puellae ante noctem   May the girls return before nightfall.
utinam nē sērō redeant puellae May the girls not return late.

Wishes for the future are expressed by the present subjunctive,
negative **nē** or **nōn**. **utinam** may be used to make it clear that the
subjunctive expresses a wish (= o that, I wish that).

3 quid faciāmus? What are we to do?

This is called the deliberative subjunctive, used when someone is
deliberating on some course of action, e.g.
quid patrī dīcāmus?   What are we to say to father?

4 velim tē adiuvāre I should like to help you.

This use, called the potential subjunctive, is commonest with **velim**
I should like, **nōlim** I should not like, **mālim** I should prefer, **ausim** I
would dare, e.g.   **nōn ausim tālia facere** I would not dare do such things.

Exercise 9.1

Translate
1 quid faciāmus? frāterculum invenīre nōn possumus.
2 domum festīnēmus et omnia parentibus nārrēmus.
3 nōlim frātrem in silvīs errantem relinquere.
4 venī. domum redeāmus ut patrem arcessāmus.
5 utinam nē īrāscātur pater. ego nōn ausim eī dīcere quid acciderit.
6 nox adest. lupōs audiō ululantēs. utinam misellus lupō nōn occurrat.
7 age, vocēmus eum rursus magnā vōce.
8 utinam nōs audiat. 'MARCE? UBI ES? VENI HUC CELERITER.'
9 ecce, misellum videō ē silvīs exeuntem. currāmus ut eum adiuvēmus.
10 Mārce, ubi fuistī? numquam tālia iterum faciās.

Exercise 9.2

Translate into Latin
1 Soldiers, the enemy are advancing quickly. What are we to do?
2 Let us not be afraid but let us fight bravely.
3 I should prefer to die in battle than betray the city.
4 Let us all run quickly to the rampart and defend the walls.
5 May we conquer the enemy and win eternal glory for ourselves.

Exercise 9.3

Read the following poem and answer the questions below
amīcus Catullī, nōmine Fabullus, Verānium in Hispāniam comitātus
erat. cum Rōmam rediisset, Catullus, quī argentō tum carēbat, eum
ad cēnam sīc invītāvit:

cēnābis bene

cēnābis bene, mī Fabulle, apud mē
paucīs, sī tibi dī favent, diēbus,
sī tēcum attuleris bonam atque magnam
cēnam, nōn sine candidā puellā
5   et vīnō et sale et omnibus cachinnīs.
haec sī, inquam, attuleris, venuste noster,
cēnābis bene; nam tuī Catullī
plēnus sacculus est arāneārum.
sed contrā accipiēs merōs amōrēs,
10 seu quid suāvius ēlegantiusve est:
nam unguentum dabo, quod meae puellae
dōnārunt Venerēs Cupīdinēsque,
quod tū cum olfaciēs, deōs rogābis,
tōtum ut tē faciant, Fabulle, nāsum.   (13)

5   **sale** salt = wit.   **cachinnīs** laughter.
6   **venuste noster** my charming friend.
8   **sacculus** my little purse.   **arāneārum** of cobwebs.
9   **contrā** in return.   **merōs amōrēs** sheer love = something you'll really
     love.
10   **seu quid . . . est** or something even more delightful and elegant.
11   **unguentum** a perfume.
12   **dōnārunt = dōnāvērunt.**   **Venerēs Cupīdinēsque** all the Loves and
     Cupids, i.e. all the powers of love (the perfume had a very erotic scent,
     like those on television advertisements).
13   **cum olfaciēs** when you smell, sniff.
14   **nāsum** nose.

1 When Fabullus comes to dinner with Catullus, what will he have to
   bring with him?
2 **arāneārum** (l.8): why cobwebs?
3 What will Fabullus get in return?
4 What will Fabullus ask the gods to do?

Compare the following translations of this poem:

1 Now, please the gods, Fabullus, you
    Shall dine here well in a day or two;
    But bring a good big dinner, mind,
    Likewise a pretty girl and wine
    And wit and jokes of every kind.
    Bring these, I say, good man, and dine
    Right well; for your Catullus' purse
    Is full – but only cobwebs bears.
    But you with love itself I'll dose,
    Or what still sweeter, finer is,
    An essence to my lady given
    By all the Loves and Venuses;
    Once sniff it, you'll petition heaven
    To make you nose and only nose.   (William Marris 1924)

2 say Fabullus
    you'll get a swell dinner at my house
    a couple three days from now (if your luck holds out)
    all you gotta do is bring the dinner
      and make it good and be sure there's plenty
    Oh yes don't forget a girl (I like blondes)
    and a bottle of wine maybe
      and any good jokes and stories you've heard
    just do that like I tell you ol' pal ol' pal
    you'll get a swell dinner
        ?
      what,
          about,
              ME?
  well;
      well here take a look in my wallet,
        yeah those're cobwebs
  but here,
        I'll give you something too
        I CAN'T GIVE YOU ANYTHING BUT
          LOVE BABY

unguentum dabo

  no?
  well here's something nicer and a little more cherce maybe
  I got a perfume see
  it was a gift to HER
    straight from VENUS and CUPID LTD.
  when you get a whiff of that you'll pray the gods
  to make you (yes you will, Fabullus)
    ALL
      NOSE   (Frank Copley 1957)

Which translation gives the sense more accurately?
Which conveys the spirit of the original better?

Exercise 9.4

Translate into Latin

1 When Catullus was living in Rome, he had many friends who wrote
  verses.
2 When he dined with Calvus, he enjoyed himself so much that he
  wanted to talk with him again.
3 He warned Calvus not to be proud and not to reject his prayers.
4 Why have you sent me such a bad book? Do you want to destroy
  your friend?
5 Since you have dared to do this, I promise that I will punish you.

and not to **nēve**   enjoy myself **mē oblectō**

Exercise 9.5

Translate into Latin

When his friend Fabullus had returned from Spain, Catullus sent him
a letter in which he asked him to dinner. But he said that he was short
of money. 'You will dine well with me,' he said, 'if you bring with you
a good big dinner and a pretty girl.' But he promised that he would
give him in return a wonderful perfume. And so Fabullus brought a
very good dinner and asked a very pretty girl to come with him. They
enjoyed themselves so much that they stayed up all night, drinking
and talking.

Spain **Hispānia, -ae,** *f.*   I am short of **careō** (2) + ablative
in return **contrā**   perfume **unguentum, -ī,** *n.*   I stay up **vigilō** (1)

plēnus sacculus est arāneārum

103

CHAPTER X

Catullus in love

necesse erat iuvenem animō tam ārdentī praeditum amōris
vulnera patī. nōn diū Rōmae habitāverat cum Clōdiae amōre
captus est. illa, mulier nōbilī genere nāta, virī cōnsulāris uxor,
fōrmā erat ēgregiā sed ingeniō prāvō et libīdinōsō. īnfēlīx
5 Catulle, nihil nisi dolōrem amor tālis tibi allātūrus erat!
  prīmum Catullus Clōdiae ipsī amōrem dēclārāre nōn audet
sed hōc carmine passerem eius salūtat amōremque suum oblīquē
significat:

passer, dēciae meae puella

  passer, dēliciae meae puellae,
10 quōcum lūdere, quem in sinū tenēre,
  cui prīmum digitum dare appetentī
  et ācrēs solet incitāre morsūs,
  cum dēsīderiō meō nitentī
  cārum nescioquid lubet iocārī,
15 et sōlāciolum suī dolōris,
  crēdō, ut tum gravis acquiēscat ārdor;
  tēcum lūdere sīcut ipsa possem
  et trīstēs animī levāre cūrās!   (2)

  mox Clōdia amōrī eius respondet; Catullus gaudiō exsultat.

20 vīvāmus, mea Lesbia, atque amēmus,
  rūmōrēsque senum sevēriōrum
  omnēs ūnius aestimēmus assis!
  sōlēs occidere et redīre possunt.
  nōbīs cum semel occidit brevis lūx,
25 nox est perpetua ūna dormienda.
  dā mī bāsia mīlle, deinde centum,
  dein mīlle altera, dein secunda centum,
  deinde usque altera mīlle, deinde centum.
  dein cum mīlia multa fēcerīmus,
30 conturbābimus illa, nē sciāmus,
  aut nē quis malus invidēre possit,
  cum tantum sciat esse bāsiōrum.   (5)

dā mī bāsia mīlle

1	**necesse erat** it was inevitable that. **praeditum** endowed with.
4	**fōrmā erat ēgregiā** was of outstanding beauty. **prāvō** vicious. **libīdinōsō** lustful.
7	**passerem eius** her sparrow. Birds were often kept as pets; we hear of pet doves, goldfinches, magpies etc. The sparrow was the bird sacred to Venus. **oblīquē** obliquely, not directly.
9	**dēliciae** (*f.pl.*) the darling.
10	**in sinū** on her lap.
11	**prīmum digitum** her finger tip.
12	**morsūs** bites, nips.
13–14	**cum . . . lubet** when it pleases (her). **dēsīderiō meō nitentī** bright-eyed with longing for me: **nitentī** (literally 'shining') agrees with 'her' (understood).
14	**cārum nescioquid iocārī** to play some sweet game.
15	**sōlāciolum** as a comfort.
16	**gravis ārdor** her fierce passion.
17	**possem** I wish I could.
18	**levāre** to relieve.

1 How does Catullus's girl play with the sparrow?
2 Why, according to Catullus, does she play with it?
3 l.15 **suī dolōris**: what pain is Catullus talking about?
4 l.16 **crēdō**: what is the significance of this aside?
5 What does Catullus wish?
6 The whole poem is addressed to the sparrow, but what do you suppose is the real subject and intention of the poem?

20	**mea Lesbia** = Clodia. Poets did not use the real name of the girls to whom they wrote their poetry but a name with the same scansion value. **Lēsbiă = Clōdiă**.
21	**rūmōrēs** gossip
22	**ūnius aestimēmus assis** let us value at one **as** = let us think not worth a penny. The **as** was a coin of very low value.
26	**mī = mihi**. **bāsia** kisses.
28	**usque** at once, without interruption.
30	**conturbābimus illa** we'll muddle the count of them.
31	**invidēre** to cast an evil eye on us. The ancients believed that evil men could cast a spell on you with their evil eye. This was particularly liable to happen if you were too happy. If Catullus and Clodia lose count of their kisses, they are less likely to suffer from the evil eye than if they know they have kissed 3,300 times.

Compare the following translations of lines 20–5.
Which do you think better conveys the feeling of Catullus's poem?

My sweetest Lesbia, let us live and love;
And though the sager sort our deeds reprove,
Let us not weigh them. Heaven's great lamps do dive
Into their west, and straight again revive.
But soon as once is set our little light,
Then must we sleep one ever-during night.   (Thomas Campion 1601)

I said to her, darling, I said
let's LIVE and
let's LOVE and
what do we care what those old
purveyors of joylessness say?
(they can go to hell, all of them)
the Sun dies every night
in the morning he's here again
you and I, now,
when our briefly tiny light flicks out,
it's night for us, one single
everlasting
Night.   (Frank Copley 1957)

    sed Clōdia mulier erat nec cōnstāns nec fidēlis; prōmittit
amōrem suum ergā Catullum perpetuum fore; Catullus autem
35 verbīs eius cōnfīdere vix potest:

iūcundum, mea vīta, mihī prōpōnis amōrem
   hunc nostrum inter nōs perpetuumque fore.
dī magnī, facite ut vērē prōmittere possit
   atque id sincērē dīcat et ex animō,
40 ut liceat nōbīs tōtā perdūcere vītā
   aeternum hoc sānctae foedus amīcitiae.   (109)

    Clōdiae persuādēre temptat ut virō relictō sibi ipsī nūbat;
scit tamen prōmissa eius inānia esse:

nūllī sē dīcit mulier mea nūbere mālle
45   quam mihi, nōn sī sē Iuppiter ipse petat.
dīcit: sed mulier cupidō quod dīcit amantī
   in ventō et rapidā scrībere oportet aquā.   (70)

    mox Clōdia alium amātōrem sūmit. Catullus dēspērat;
animus eius mōtibus contrāriīs distrahitur.

50 ōdī et amō. quārē id faciam, fortasse requīris.
   nescio, sed fierī sentiō et excrucior.   (85)

106

ō dī, mē miserum aspicite

34   **amōrem ergā** love for.

36   **mea vīta** my life = my darling.   **prōpōnis** you offer, promise. This is followed by the accusative and infinitive: subject **amōrem hunc nostrum.**

39   **sincērē** sincerely.

40   **tōtā vītā = tōtam vītam** through our whole life.

46   **sed . . . amantī = quod** (what) **mulier dīcit cupidō** (eager) **amantī.**

47   **rapidā** running. Compare this line with:
       Woman's faith and woman's trust –
       Write the characters in dust,
       Stamp them on the running stream,
       Print them on the moonlight's beam.   (Sir Walter Scott)

49   **mōtibus** by emotions.   **distrahitur** is torn apart.

50   **quārē** why?

51   **excrucior** I am in torture.

ōdī et amō . . . these two lines form a complete poem which has been much admired. What makes them so powerful?

cōnātur Clōdiae amōrem pectore ēicere sed nōn potest; sīc
sē ipsum hortātur ut obdūret.

    miser Catulle, dēsinās ineptīre,
55   et quod vidēs perīsse perditum dūcās.
    fulsēre quondam candidī tibī sōlēs,
    cum ventitābās quō puella dūcēbat
    amāta nōbīs quantum amābitur nūlla;
    ibi illa multa cum iocōsa fīēbant
60   quae tū volēbās nec puella nōlēbat,
    fulsēre vērē candidī tibi sōlēs.
    nunc iam illa nōn volt; tū quoque impotēns nōlī,
    nec quae fugit sectāre, nec miser vīve,
    sed obstinātā mente perfer, obdūrā.
65   valē, puella. iam Catullus obdūrat,
    nec tē requīret nec rogābit invītam.
    at tū dolēbis, cum rogāberis nūlla.
    scelesta, vae tē, quae tibī manet vīta?
    quis nunc tē adībit? cui vidēberis bella?
70   quem nunc amābis? cuius esse dīcēris?
    quem bāsiābis? cui labella mordēbis?
    at tū, Catulle, dēstinātus obdūrā.   (8)

tandem amor ille quō nūper gāvīsus erat morbus animī
factus est; deōs ōrat ut hanc pestem perniciemque sibi ēripiant:

75   sī qua recordantī benefacta priōra voluptās
        est hominī, cum sē cōgitat esse pium,
    nec sānctam violāsse fidem, nec foedere nūllō
        dīvum ad fallendōs nūmine abūsum hominēs,
    multa parāta manent in longā aetāte, Catulle,
80   ex hōc ingrātō gaudia amōre tibī.
    nam quaecumque hominēs bene cuiquam aut dīcere possunt
        aut facere, haec ā tē dictaque factaque sunt.
    omnia quae ingrātae periērunt crēdita mentī.
        quārē iam tē cūr amplius excruciēs?
85   quīn tū animō offirmās atque istinc tēque redūcis,
        et dīs invītīs dēsinis esse miser?
    difficile est longum subitō dēpōnere amōrem.
        difficile est, vērum hoc quālubet efficiās;
    ūna salūs haec est, hoc est tibi pervincendum,
90   hoc faciās, sīve id nōn pote sīve pote.

52	**Clōdiae amōrem** his love for Clodia.
53	**obdūret** harden his heart.
54–5	**dēsinās . . . dūcās** cease . . . consider (jussive subjunctives instead of imperatives). **ineptīre** to play the fool.
56	**fulsēre = fulsērunt** shone.
57	**cum ventitābās** when you used to go.
58	**quantum nūlla** as (much as) no girl . . . .
59	**ibi** then. **iocōsa** jolly, merry.
62	**volt = vult. impotēns** weakling.
63	**quae** (a girl) who. **sectāre**: imperative of **sector -ārī** I follow.
68	**scelesta, vae tē** poor girl, alas for you.
70	**cuius esse dīcēris** whose (girl) will you be said to be?
71	**bāsiābis** will you kiss? **cui labella mordēbis** whose lips will you bite?
72	**dēstinātus** resolute.

1 How is poem 8 constructed – i.e. into what blocks of sense does it divide? The clue is given by the persons of the verbs – to whom is Catullus speaking in each section?
2 What is the tone of each section?
3 At the end of the poem, does it seem to you that Catullus has succeeded in hardening his heart?

74	**pestem perniciemque** plague and destruction. **sibi** from him.
75–6	**sī qua . . . voluptās est** if there is any pleasure . . . .
77	**violāsse = violāvisse; violāre =** to violate, break. **nec foedere nūllō** and in no pact.
78	**dīvum nūmine abūsum (esse)** he has abused the power of the gods, i.e. by swearing an oath by the gods and then breaking it.
83	**crēdita** entrusted to.
84	**quārē** and so. **tē amplius excruciēs** should you torture yourself any more?
85	**quīn** why not? **animō offirmās** be firm of heart. **istinc** from there = from your love. **-que** both (both withdraw and cease).
88	**vērum** but. **quālubet** somehow or other.
89	**pervincendum** must be won, achieved.
90	**pote** possible.

ō dī, sī vestrum est miserērī, aut sī quibus umquam
    extrēmam iam ipsā in morte tulistis opem,
mē miserum aspicite et, sī vītam pūriter ēgī,
    ēripite hanc pestem perniciemque mihi,
95   quae mihi subrēpēns īmōs ut torpor in artūs
    expulit ex omnī pectore laetitiās.
nōn iam illud quaerō, contrā mē ut dīligat illa,
    aut, quod nōn potis est, esse pudīca velit.
ipse valēre optō et taetrum hunc dēpōnere morbum.
100   ō dī, reddite mī hoc prō pietāte meā.   (76)

1 The poem seems to be constructed in three blocks of sense; in the
   middle block, Catullus is arguing with himself. Summarize the
   meaning of each block and say what the tone of each is. Where do
   you feel the emotion becomes strongest?
2 Catullus says that he has been **pius**; **pius** means loyal, or devoted, to
   the gods, one's loved ones and one's country. How has he shown
   pietās? How did Clodia fail to show it?
3 After reading this poem how do you feel towards Catullus? Can you
   imagine that you might ever experience similar emotions yourself?

V	
aestimō (1)	I value
amplius (adverb)	more; longer
fōrma, -ae, *f.*	shape, beauty
ingrātus-a-um	ungrateful
invītus-a-um	unwilling
levis, -e	light
levō (1)	I lighten
lūgeō, -ēre, lūxī	I mourn, grieve
nūbō, -ere, nupsī, nuptum +dative	I marry
ops, opis, *f.*	help
pūrus-a-um	pure
prior, prius	former
prius (adverb)	before
priusquam (conjunction)	before
prō + ablative	in front of; on behalf of; instead of
requīrō, -ere, requīsīvī, requīsītum	I seek for, ask, ask for
sīve … sīve …	whether … or

110

91	**ō dī, sī vestrum est miserērī** oh, gods, if it is in you to feel pity.
95	**subrēpēns** creeping. **ut torpor** like a paralysis. **īmōs in artūs** into the depths of my limbs.
97	**contrā** in return.
98	**quod nōn potis est** which is not possible. **pudīca** chaste.
99	**taetrum** foul.
100	**reddite** grant. **prō** in return for.

🄴 Word building

Make sure that you know the following compounds of **cēdō, cēdere, cessī, cessum**

accēdō	I approach; I am added	**incēdō**	I enter
antecēdō	I go before, excel	**intercēdō**	I intervene
concēdō	I withdraw, yield	**praecēdō**	I go before, excel
dēcēdō	I go away, depart	**prōcēdō**	I go forwards, advance
discēdō	I leave	**recēdō**	I go back, retire
excēdō	I withdraw	**succēdō**	I succeed, come after

Give English derivatives from as many of these compounds as you can.

🄵 Passive imperatives

1	*2*	*3*	*4*	*5*
amāre	monēre	regere	audīre	capere
amāminī	monēminī	regiminī	audīminī	capiminī

These forms are most common from deponent verbs.

Translate
1 fīlī, cōnāre mē adiuvāre.
2 amīcī, sequiminī mē in oppidum.
3 mīlitēs, fortiter in hostēs prōgrediminī.

4 statim proficīscere, amīce, ad canem quaerendum.
5 nōlī tam sevērus esse, amīce; meliōra loquere.

🄶 Price and value

1 Price

hanc vīllam centum mīlibus sestertium ēmī
I bought this house for 100,000 sesterces.

If an exact price is stated, the ablative is used.

2 Value

hanc vīllam magnī aestimō I value this house highly.
haec vīlla parvī est This house is worth little.

Value is expressed by the genitive case.

Exercise 10.1

Translate

1 vīsne tū hunc equum mihi vendere?
2 equum tibi vendere volō mīlle sestertiīs.
3 nimis rogās. equus nōn tantī est.
4 errās, amīce. equum plūrimī aestimō.
5 quīngentīs sestertiīs eum emere volō.

6 nōlī nūgās nārrāre. licet tibi eum emere quīnquāgintā et septingentīs sestertiīs. placetne tibi?
7 mihi placet, quamquam multō minōris est.

Exercise 10.2

Translate

1 sī festīnāvissēs, fortasse amīcum tuum servāvissēs.
2 nōnne tē pudet amīcum prōdere?
3 statim tibi proficīscendum est ut eum quaerās.
4 servōs praemitte, quī eī dīcant tē mox adfore.
5 age, tē comitābor. statim proficīscāmur.
6 sī iter iam nunc inierimus, eum in Italiā assequēmur.
7 utinam nē sērō adveniāmus.
8 es vir summā prudentiā. cōnsilium tuum magnī aestimō.
9 tēcum iter faciam ut amicō auxiliō sīmus.
10 omnia paranda sunt quae ad tantum iter ūsuī erunt.

Exercise 10.3

Translate the following poem and answer the questions below

A lament for Lesbia's sparrow

lūgēte, ō Venerēs Cupīdinēsque,
et quantum est hominum venustiōrum;
passer mortuus est meae puellae,
passer, dēliciae meae puellae,
5   quem plūs illa oculīs suīs amābat;
nam mellītus erat suamque nōrat
ipsam tam bene quam puella mātrem,
nec sēsē ē gremiō illius movēbat,
sed circumsiliēns modo hūc modo illūc
10 ad sōlam dominam usque pīpiābat:
quī nunc it per iter tenēbricōsum
illud, unde negant redīre quemquam.
at vōbīs male sit, malae tenēbrae
Orcī, quae omnia bella dēvorātis.
15 tam bellum mihi passerem abstulistis.
ō factum male! ō miselle passer!
tuā nunc operā meae puellae
flendō turgidulī rubent ocellī.   (3)

2	**et quantum est** and whatever there are of men of finer feeling = all men of finer feeling.

2   **et quantum est** and whatever there are of men of finer feeling = all men of finer feeling.

6   **mellītus** honey sweet, a honey.   **nōrat = nōverat** knew.

6–7   **suam ipsam** his mistress (compare Irish: 'himself is out').

8–9   **gremiō** lap.   **circumsiliēns** hopping round.

10   **usque pīpiābat** chirped continually, kept chirping.

11   **tenēbricōsum** shadowy.

13   **vōbīs male sit** may it be bad for you = curse on you.

14   **Orcī** of hell.

17   **tuā nunc operā** through your fault.

18   **turgidulī** swollen.   **rubent** are red.

1 How did the sparrow show its affection for its mistress?

2 Where is it going now?

3 Why does Catullus curse the shades of Orcus?

4 Read l.11 aloud several times and listen to the sounds and rhythm. How do these help to give a picture of the sparrow's journey?

5 The poem is a lament for the death of Lesbia's sparrow, but what else does Catullus express in it? Quote the lines which support your view.

Exercise 10.4

Translate into Latin

1 When Sabinus was collecting corn for (**in** + accusative) the winter, the Gauls suddenly attacked his camp.

2 The Romans defended the camp so bravely that the Gauls could not take it.

3 Then the chief of the Gauls asked Sabinus to send representatives (**lēgātī, -ōrum,** *m.*) to talk about conditions.

4 He promised that he would give the Romans a safe journey through his territory.

5 If Sabinus had not accepted these conditions, he would not have destroyed his legion.

Exercise 10.5

Translate into Latin

Sabinus persuaded the council of war (**concilium, -ī,** *n.*) to accept the conditions which Ambiorix had proposed. And so the Romans, having left the camp at dawn, marched in a long column towards the woods. But the Gauls realised what the Romans had in mind and, when they knew that they (i.e. the Romans) were about to set out, they placed an ambush in the woods and waited for their arrival. When the greater part of the Roman column had descended into a large valley, the Gauls suddenly showed themselves and began to join battle.

For those who have the time and inclination to read what
happened to Clodia in the end.

is quī Clōdiam Catullō ēripuerat, erat M.Caelius Rūfus, amīcus
et Catullī et Cicerōnis, cuius litterās ad Cicerōnem scrīptās iam
lēgistī. cum ille amīcum suum sīc prōdidisset, Catullus hīs
versibus eum vituperāvit:

5   Rūfe, mihī frūstrā ac nēquīquam crēdite amīce
       (frūstrā? immō magnō cum pretiō atque malō),
    sīcine subrēpstī mī atque intestīna perūrēns
       ei miserō ēripuistī omnia nostra bona?
    ēripuistī, heu heu, nostrae crūdēle venēnum
10   vītae, heu heu, nostrae pestis amīcitiae.   (77)

    Caelius tamen ingeniō fuit dūriōre quam Catullus et
fortiōre; brevī tempore, Clōdiae animō perspectō, eam reiēcit;
quae īrā impotentī ārdēns poenās eī sūmere cōnāta est; nam
lītem in eum intendit. frāter eius, Clōdius, Caelium accūsat,
15 Cicerō autem dēfendit. tandem et Caelius absolūtus est et Clōdia
ā Cicerōne tam asperē tractāta est ut ab omnibus contempta
Rōmā discesserit neque quicquam dē eā posteā audīvimus.
    in ōrātiōne prō Caeliō sīc in eam invehitur Cicerō:

    sī quae nōn nūpta mulier domum suam patefēcerit omnium
20 cupiditātī palamque sēsē in meretrīciā vītā conlocārit, sī hoc in
urbe, sī in hortīs, sī in Bāiārum illā celebritāte faciat, sī dēnique
ita sēsē gerat nōn incessū sōlum sed ōrnātū et comitātū, nōn
lībertāte sermōnum, sed etiam complexū, ōsculātiōne, āctīs,
nāvigātiōne, convīviīs, ut nōn sōlum meretrīx sed etiam proterva
25 meretrīx videātur: cum hāc sī quī adulēscēns forte fuerit, utrum
hic tibi adulter an amātor videātur? oblīvīscor iam iniūriās tuās,
Clōdia, dēpōnō memoriam dolōris meī; quae abs tē crūdeliter in
meōs mē absente facta sunt neglegō; nē sint haec in tē dicta quae
dīxī. sed ex ipsā tē requīrō . . . sī quae mulier sit eius modī
30 quālem ego paulō ante dēscrīpsī, tuī dissimilis, vītā īnstitūtōque
meretrīciō, cum hāc aliquid adulēscentem hominem habuisse
ratiōnis num tibi perturpe aut perflāgitiōsum esse videātur?

In this case Cicero, the model of respectability, was in the awkward
position of having to defend the character of a young man who had
certainly sown a lot of wild oats. He does this with great skill and
humour and his attack on Clodia is devastating. The acquittal of
Caelius was equivalent to the condemnation of Clodia; she
disappeared from the scene without further trace.

1	**Caelius Rūfus**: see Cicero, Chapter 7, page 76. Catullus calls him by his last name, **Rūfus**, in this poem.
5	**mihī crēdite** trusted by me.
6	**immō** no, rather. . . . **magnō cum pretiō** at a heavy price.
7	**sīcine = sīc-ne. subrēpstī** have you crept up on me? **intestīna perūrēns** burning my bowels, i.e. my inmost heart.
8	**ei:** an exclamation of pain 'oh!'. **miserō** from unhappy me = from me in my misery. **nostra = mea**.
9	**nostrae crūdēle venēnum** the cruel poison of my life.
13	**īrā impotentī ārdēns** burning with uncontrollable anger.
14	**lītem in eum intendit** brought a law suit against him. This case was a *cause célèbre*. Clodius led the prosecution, Cicero the defence; Caelius also spoke in his own defence. The charges, according to Cicero, were very flimsy, and the real object of the prosecution was to discredit Caelius's character. Cicero's defence was largely devoted to destroying Clodia's character and discrediting her evidence. The prosecution, and indeed the whole court, must have been aghast when Cicero launched into this attack, in which he openly called one of the leading ladies of Roman society a shameless whore.
19	**nōn nūpta**: Clodia's husband was now dead. **patefēcerit** opened.
20	**cupiditātī** to the lust. **palamque . . . conlocārit = conlocāverit** openly established herself in the life of a whore.
21	**hortīs** the public gardens. **in Bāiārum illā celebritāte** in all those crowds at Baiae. Baiae was a fashionable seaside resort on the bay of Naples.
22	**nōn incessū sōlum** not by her gait (i.e. the way she walked) only
23	**ōsculātiōne** by her manner of kissing. **āctīs** beach parties.
24	**nāvigātiōne** sailing parties. **meretrīx** whore. **proterva** wanton, lascivious.
26	**adulter . . . amātor**: to be an adulter was disreputable, to be an **amātor** quite respectable. **iniūriās tuās** the wrongs you did me, i.e. the wrongs done to Cicero and his family (**in meōs**) at the time of his exile.
28–9	**nē sint dicta** . . . let these things which I said not have been said against you = let's suppose I did not refer to you when I said this.
30–1	**vītā īnstitūtōque meretrīciō** by her set way of life a whore.
31–2	**aliquid ratiōnis** some dealings.
32	**perturpe aut perflāgitiōsum esse videātur** does it seem absolutely disgraceful and scandalous that . . . .

FACITIS · VOBIS · SVAVITER · EGO · CANTO · EST · ITA · VALEAS

CHAPTER XI

Catullus – the sequel

Catullus cum aliquid recreātus esset, duōs amīcōs rogāvit ut
nūntium suum ultimum ad Clōdiam ferrent:

Fūrī et Aurēlī comitēs Catullī,
sīve in extrēmōs penetrābit Indōs,
5   lītus ut longē resonante Eōā
    tunditur undā,

sīve in Hyrcānōs Arabasve mollēs,
seu Sagās sagittiferōsve Parthōs,
sīve quae septemgeminus colōrat
10   aequora Nīlus,

sīve trāns altās gradiētur Alpēs,
Caesaris vīsēns monimenta magnī,
Gallicum Rhēnum horribilēsque ulti-
    mōsque Britannōs,

15   omnia haec, quaecumque feret voluntās
caelitum, temptāre simul parātī,
pauca nūntiāte meae puellae
    nōn bona dicta.

cum suīs vīvat valeatque moechīs,
20   quōs simul complexa tenet trecentōs,
nūllum amāns vērē, sed identidem omnium
    īlia rumpēns;

nec meum respectet, ut ante, amōrem,
quī illius culpā cecidit velut prātī
25   ultimī flōs, praetereunte postquam
    tāctus arātrō est.

(11)

1 This poem is constructed in three parts. Where would you make the
  divisions and what is the tone of each part?
2 1.18 **nōn bona dicta**: why do these words give the reader a shock?
3 The first twelve lines of the poem give a list of dangerous missions
  on which Furius and Aurelius would accompany Catullus. But there
  is one more mission yet more dangerous; what is it?
4 There are some striking sound effects in lines 5–6 (**lītus. . .undā**)
  and again in lines 19–20; can you spot them?
  How do they serve in each case to reinforce the sense?
5 What would you say was Catullus's state of mind when he wrote this
  poem? How did he feel towards Lesbia and towards their past love?

1	**aliquid recreātus esset** was somewhat recovered.
3	**Fūrī et Aurēlī comitēs Catullī** Furius and Aurelius, (who will be) companions of Catullus . . . , whether he goes to the remotest East (**extrēmōs Indōs**), or to the Hyrcānī (who lived near the Caspian Sea), or the Arabs, or the Sagae (nomads on the Northern borders of Persia), or to the Parthians or to Egypt, or whether he crosses the Alps to Gaul and the Rhine and on to the horrible Britons at the end of the world.
5–6	**ut** where. **longē . . . undā** is pounded by the far sounding eastern wave(s).
8	**sagittiferōs** arrow-bearing. The Parthians were famous for their mounted archers.
9	**septemgeminus** seven-mouthed. The Nile colours the waters of the sea with the silt it carries down.
12	**Caesaris monimenta magnī:** anyone travelling over the Alps to Gaul and the Rhine at this time would be bound to think of Caesar's campaigns. In fact, this poem may have been written in 55 BC, the year Caesar crossed the Rhine and invaded Britain.
16	**caelitum** of the heavenly ones = the gods. **simul** together (with me).
19	**moechīs** adulterers. **valeat** let her flourish = good luck to her! (In what tone is this said?)
20	**quōs simul . . . tenet trecentōs** whom she holds 300 at a time. (**trecentōs** is used of an indefinitely large number.)
21	**identidem** again and again.
22	**īlia rumpēns** bursting the balls.
23	**nec . . . respectet** and let her not look for.
24–5	**prātī ultimī** on the edge of the meadow.
26	**arātrō** by the plough.

sagittifer Parthus

iam cōnfirmātō animō cōnstituit Rōmā relictā cohortī
praetōris Bīthȳniae sē iungere. tōtum annum in Bīthȳniā
manendum erat; multōs amīcōs ibi fēcit sed nec praetor nec locus
30 nec caelum eī placēbat. tandem discēdere licuit. amīcōs quōs in
Bīthȳniā fēcerat sīc valēre iubet:

iam vēr ēgelidōs refert tepōrēs,
iam caelī furor aequinoctiālis
iūcundīs Zephyrī silēscit aurīs.
35 linquantur Phrygiī, Catulle, campī
Nīcaeaeque ager ūber aestuōsae.
ad clārās Asiae volēmus urbēs.
iam mēns praetrepidāns avet vagārī,
iam laetī studiō pedēs vigēscunt.
40 ō dulcēs comitum valēte coetūs,
longē quōs simul ā domō profectōs
dīversae variē viae reportant.   (46)

ad clārās Asiae volēmus urbēs   Ephesus

sīc Catullus Bīthȳniā laetissimus discessit clārāsque Asiae
urbēs vīsit. iter et longum et iūcundum in phasēlō fēcit quō
45 maximē gaudēbat. cum tandem domum rediisset, convīvium
amīcīs ōrnāvit apud Sirmiōnem. hospitēs dūxit ut phasēlum illum
vidērent, quī nunc in lacū Bēnācō prope vīllam Catullī
reconditus erat:

118

27–8	**cohortī praetōris Bīthȳniae** to the staff of the governor of Bithynia. Young men of good family were taken on the staff of provincial governors as aides-de-camp; it gave them useful provincial experience. While Catullus was in Bithynia, he wrote a poem to Veranius and Fabullus, who were now serving in the **cohors** of the governor of Macedonia, to ask how they were getting on. Bithynia was on the south coast of the Black Sea.
30	**caelum** the climate.
32	**ēgelidōs** not cold, no longer chill. **tepōrēs** warmths = warm days.
33	**caelī ... aequinoctiālis:** the vernal equinox (when night and day are exactly the same length) falls on 22 March; at this time there are commonly furious gales.
34	**Zephyrī:** Zephyr is the warm west wind, which heralds the coming of spring. **silēscit** quietens down.
35	**linquantur = relinquantur.** **Phrygiī:** Phrygia was the ancient name for Bithynia.
36	**Nīcaeae:** Nicaea was the capital of Bithynia. **über** fertile. **aestuōsae** sweltering.
37	**clārās:** the famous cities of Asia would include Troy, Miletus, Ephesus and Rhodes.
38	**praetrepidāns** trembling in anticipation. **avet** longs. **vagārī** to be off.
40	**coetūs** companies, bands.
42	**dīversae** different. **variē** variously = by various routes.

1 What was the weather like when Catullus wrote this poem?
2 What does Catullus intend to do?
3 What are his feelings about it all? Have you ever felt like this?
4 What does he say about his friends' journey from home and about their return journey?
5 The poem is divided into two halves, joined by a linking line. Explain this structure. The opening lines of the second half contain sound echoes of the first two lines; can you spot them? What link is there between the season and Catullus's feelings?

44	**phasēlō** yacht.
45	**convīvium** a party.
46	**hospitēs** guests.
48	**reconditus** laid up.

phasēlus ille

phasēlus ille quem vidētis, hospitēs,
50 ait fuisse nāvium celerrimus,
neque ūllius natantis impetum trabis
nequīsse praeterīre, sīve palmulīs
opus foret volāre sīve linteō.
et hoc negat minācis Hadriāticī
55 negāre lītus īnsulāsve Cycladas
Rhodumque nōbilem horridamque Thrāciam
Propontidā trucemve Ponticum sinum
(ubi iste post phasēlus anteā fuit
comāta silva – nam Cytōriō in iugō
60 loquente saepe sībilum ēdidit comā).
Amastri Pontica et Cytōre buxifer,
tibi haec fuisse et esse cognītissima
ait phasēlus: ultimā ex orīgine
tuō stetisse dīcit in cacūmine,
65 tuō imbuisse palmulās in aequore;
et inde tot per impotentia freta
erum tulisse, laevā sīve dexterā
vocāret aura, sīve utrumque Iūppiter
simul secundus incidisset in pedem;
70 neque ūlla vōta lītorālibus deīs
sibi esse facta, cum venīret ā marī
novissimō hunc ad usque limpidum lacum.
sed haec prius fuēre; nunc recondita
senet quiēte sēque dēdicat tibi,
75 gemelle Castor et gemelle Castoris.   (4)

route of Catullus's yacht

Rōmam mox rediit amīcōsque veterēs revīsit. quondam
Vārus eī in forō occurrit ōtiōsō, quī dūxit eum ad puellam suam
vīsendam. dē hāc rē Catullus fābulam fēcit ex quā appāret eum,
quamvīs mala passus esset, facētiās certē nōn perdidisse.

80 Vārus mē meus ad suōs amōrēs
vīsum dūxerat ē forō ōtiōsum –
scortillum (ut mihi tum repente vīsum est)
nōn sānē illepidum neque invenustum;
hūc ut vēnimus, incidēre nōbīs
85 sermōnēs variī, in quibus, quid esset
iam Bīthȳnia, quō modō sē habēret,
et quōnam mihi prōfuisset aere.

50	**ait fuisse celerrimus** says it was fastest.
51–2	**neque ūllius . . . praeterīre** nor was it unable (= and it could) overtake the speed of any ship (**trabis**) afloat.
52	**palmulīs** with oars
53	**linteō** with canvas.
54	The yacht continues to tell its own story, tracing its journey backwards from Lake Benacus (Sirmio) to its place of origin (Bithynia); see map. The passage starts with another double negative expression: 'it denies that the shore of the threatening Adriatic sea denies this . . .' = 'the yacht says that the shore of the Adriatic cannot deny this'.
57	**trucemve Ponticum sinum** or the savage Pontic gulf = the savage Black Sea.
58	**iste post phasēlus** it, afterwards a yacht = the yacht to be.
59	**comāta silva** a leafy forest. **Cytōriō in iugō** on the hills of Cytorus. Cytorus and Amastris(l.61) were ports on the Black Sea.
60	**sībilum** a whisper, whistle. The tree speaks when the wind whistles through its foliage (**comā**).
61	**Amastri . . . Cytōre:** vocatives. The yacht calls on them to bear her witness. **buxifer** box-bearing. Box wood was used in building ships.
64	**tuō in cacūmine** on your peak(s).
65	**imbuisse palmulās** (first) dipped its oars.
66	**impotentia freta** raging seas.
67	**erum** its master. **laevā sīve dexterā** = sīve laevā sīve dexterā whether from the left or from the right = from port or starboard.
68–9	**Iuppiter secundus** a following Jupiter. Iuppiter, god of the sky, here = wind. **utrumque in pedem** on both feet (of the sail): **pēs** foot or sheet, one on each corner of the bottom of the square sail.
70	**neque ūlla vōta : vōta** vows. Vows would be made to the gods of the shore in times of danger, e.g. 'If you, gods, bring me safe to shore, I will offer you a sacrifice.' Catullus's yacht had never had to make such vows.
71	**sibi** by it (dative of the agent).
72	**novissimō** latest. Its latest sea was the Adriatic, where it had made its last voyage.
73	**sed haec prius fuēre** but this was in the past. **recondita** laid up; it is best to take this as nominative feminine, agreeing with the yacht (**nāvis**).
74	**senet** grows old
75	**gemelle Castor** twin Castor. Castor and Pollux, the heavenly twins, protected seafarers and ships.
77	**ōtiōsō** at leisure.
78	**appāret** it is plain that . . . .
79	**quamvīs mala passus esset** however bad the things he had suffered. **facētiās** his wit.
80	**amōrēs** his love, his girl friend.
81	**vīsum** to see. (**vīsum** is the supine, expressing purpose.)
82	**scortillum** a little tart. **tum repente** at the actual time.
83	**nōn sānē illepidum neque invenustum** . . . certainly not without charm and grace.
84	**incidēre** = incidērunt there cropped up.
85–6	**quid esset iam Bīthȳnia** what Bithynia was like now.
86	**quō modō sē habēret** how it was getting on.
87	**quōnam . . . aere** by what money it had benefited me = how much money I had made out of it. Romans serving in the provinces expected to come back with their pockets well lined.

respondī id quod erat – nihil neque ipsīs
nec praetōribus esse nec cohortī,
90 cūr quisquam caput unctius referret –
praesertim quibus esset irrumātor
praetor, nec faceret pilī cohortem.
'at certē tamen,' inquiunt 'quod illīc
nātum dīcitur esse, comparāstī
95 ad lectīcam hominēs.' ego (ut puellae
ūnum mē facerem beātiōrem)
'nōn' inquam 'mihi tam fuit malignē,
ut, prōvincia quod mala incidisset,
nōn possem octō hominēs parāre rēctōs.'
100 (at mī nūllus erat nec hīc neque illīc,
frāctum quī veteris pedem grabātī
in collō sibi collocāre posset.)
hīc illa, ut decuit cinaediōrem,
'quaesō,' inquit 'mihi, mī Catulle, paulum
105 istōs commodā; nam volō ad Serāpim
dēferrī,' 'manē,' inquiī puellae,
'istud quod modo dīxeram mē habēre. . .
fūgit me ratiō: meus sodālis –
Cinna est Gāius – is sibī parāvit.
110 vērum utrum illius an meī, quid ad mē?
ūtor tam bene quam mihī parārim –
sed tū īnsulsa male et molesta vīvis,
per quam nōn licet esse neglegentem!'   (10)

Serāpis

        paulō post periit, adhūc iuvenis. quis scit an Clōdiae
infidēlitās eum ad mortem immātūram impulerit?

Ⓥ

adimō, adimere, adēmī, adēmptum	I take away
aequor, aequoris, *n.*	sea
aes, aeris, *n.*	bronze; money
aura, -ae, *f.*	breeze, air
cinis, cineris, *m./f.*	ash
complector, complectī, complexus	I embrace
complexus, -ūs, *m.*	embrace
convīvium, -ī, *n.*	dinner party
mollis, -e	soft
mūnus, mūneris, *n.*	gift; duty
ōrnō (1)	I equip, provide, adorn
parō (1)	I prepare; I get, acquire
varius-a-um	various, different
vōtum, -ī, *n.*	vow

88–9	**id quod erat** that which was = the facts. **nihil neque ipsīs . . . esse . . .** there was nothing for the natives themselves nor the governors nor his staff.
90	**cūr** (no reason) why. **quisquam** anyone **caput unctius referret** should bring back his head better oiled = come back any richer. Romans frequently anointed themselves with perfume, especially at dinner parties; these perfumes were expensive. After being in Bithynia, you would not be in a position to buy better perfumes.
91	**praesertim** especially. **irrumātor** a bugger.
92	**nec faceret pilī** and did not value his staff at a hair = didn't give a hair for his staff.
93	**at certē** but at least.
93–4	**quod illīc nātum esse dīcitur** what is said to be born there = what is said to be the native product.
94	**comparāstī** = **comparāvistī**.
95	**ad lectīcam hominēs** litter men = men to carry a litter.
95–6	**ut . . . beātiōrem** to make myself out to be the one luckier man = to make myself seem exceptionally lucky.
97	**nōn . . . malignē** I didn't do so badly.
99	**hominēs rēctōs** straight-backed men. It took eight men to carry a litter.
100	**at mī nūllus erat** but I had none.
101	**grabātī** of a camp bed.
102	**collō** on his neck.
103	**hīc** at this point. **ut decuit cinaediōrem** as suited a real whore = as you would expect from . . . .
104	**quaesō** = **quaerō** I ask, please.
105	**commodā** lend. **ad Serāpim** to the temple of Serapis. Serapis was an Egyptian god, whose cult was popular at Rome at this time.
106	**inquiī** I said. What Catullus says in these lines is pretty incoherent; he is confused and indignant at being caught out.
107	**modo** just now.
108	**fūgit mē ratiō** reason escaped me = I made a mistake. **sodālis** friend.
109	**Cinna est Gāius** it's Cinna, that is Gaius Cinna.
110	**vērum . . . meī** but whether they are his or mine. **quid ad mē?** what does it matter to me?
111	**parārim** = **parāverim**.
112	**sed tū . . . vīvis** but you are a very silly and boring girl: **vīvis** almost = **es**.
113	**per quam . . . neglegentem** through whom one is not allowed to be careless = who won't let a man make a mistake.

Word building

Make sure that you know the following compounds of
pōnō, pōnere, posuī, positum

antepōnō	I put before, prefer	**oppōnō**	I oppose
compōnō	I arrange, settle	**praepōnō**	I put (someone) in command of
dēpōnō	I put down, put away	(+ acc. and dat.)	(someone or something)
dispōnō	I arrange	**prōpōnō**	I put forward, propose
expōnō	I put out, disembark; I explain	**repōnō**	I put back, put aside, deposit
impōnō	I put in, embark		

Ⓖ Some uses of the dative case

1 Dative verbs revised

The following are the commonest verbs taking the dative:

crēdere*	to believe, entrust	**nocēre**	to harm
favēre	to favour, support	**nūbere**	to marry
fīdere ⎫ **cōnfīdere** ⎬	to trust	**pārēre**	to obey
		parcere	to spare
ignōscere	to pardon	**placēre**	to please
imperāre*	to order	**studēre**	to study
invidēre	to envy	**suādēre** ⎫ **persuādēre** ⎬	to persuade
īrāscī	to grow angry with		
minārī*	to threaten		

*These verbs can also take a direct object, e.g. **hoc tibi imperō**.

occurrere	to meet
succurrere	to run to help
resistere	to resist

and other compounds of intransitive verbs.

Compounds of 'sum'

adesse	to be present
dēesse	to be lacking, fail
praeesse	to be in charge of
prōdesse	to benefit

NB Dative verbs are used impersonally in the passive,
e.g. **nōbīs imperātum est** it was ordered to us = we were
ordered.

Exercise 11.1

Translate

1 mihi nōn crēditum est.
2 captīvīs parcētur.
3 nōbīs persuāsum est ut domum redīrēmus.
4 hostibus fortiter resistitur.
5 omnibus ignōtum est.
6 mīlitibus imperātum est ut in statiōne manērent.

7 The king will be obeyed.
8 We have not been harmed.
9 The boy was ordered to wait in the town.
10 I was persuaded to help the old man.

2 The dative of the person concerned: a wide range of uses

nōmen iuvenī est Horātius	The young man's name is Horatius.
est mihi gladius	I have a sword
cui labella mordēbis?	Whose lips will you bite?
gladium tibi adimam	I will take the sword from you.

Exercise 11.2

Translate

1 nōbīs cum semel occidit brevis lūx, nox est perpetua ūna dormienda.
2 ea causa mihi in Asiam proficīscendī fuit.
3 mihi omnis sermō est cum litterīs.
4 ēripite hanc pestem mihi.
5 fortūna mihi tē abstulit.
6 voluptās est hominī recordantī priōra benefacta.

7 Cicerō prō certō habēbat sē Antōniō ēripī non posse.
8 nōbīs* statim redeundum est.
9 omnia Caesarī* simul agenda erant.
10 phasēlus negat ūlla vōta deīs sibi* facta esse.

*These are examples of the so-called dative of the agent, where you would expect **ā** + ablative.

Exercise 11.3

Translate

1 cum trīclīnium intrāre cōnārēmur, servus accurrit et nōbīs prōcubuit ad pedēs.
2 nōs rogāre coepit ut sē poenae ēriperet.
3 dīxit subducta esse sibi vestīmenta dominī in balneīs, quae vix fuissent decem sestertiōrum.
4 in ātrium regressī dominum rogāvimus ut servō remitteret poenam.
5 ille 'erant mihi' inquit 'optima vestīmenta. sed cum veniam prō istō servō pessimō petātis, poenam eī remittam.'

Exercise 11.4

Translate and answer the questions below

Elegy on the death of his brother

(Catullus's brother had died near Troy; Catullus visited his tomb on his way to or from Bithynia.)

<blockquote>

multās per gentēs et multa per aequora vectus
    adveniō hās miserās, frāter, ad īnferiās,
ut tē postrēmō dōnārem mūnere mortis
    et mūtam nēquīquam alloquerer cinerem.

5   quandoquidem fortūna mihī tētē abstulit ipsum,
    heu miser indignē frāter adēmpte mihi,
nunc tamen intereā haec, prīscō quae mōre parentum
    trādita sunt trīstī mūnere ad īnferiās,
accipe frāternō multum mānantia flētū,

10   atque in perpetuum, frāter, avē atque valē.   (101)

</blockquote>

1	**vectus** carried = sailing, travelling.
2	**īnferiās** funeral offerings.
3	**ut tē dōnārem** to present you with. **mūnere mortis** the gift due to death.
4	**mūtam** dumb. **nēquīquam** in vain.
5	**quandoquidem** since. **tētē = tē**.
6	**indignē** unworthily, cruelly.
7–8	**haec** these (gifts). The traditional gifts to the dead were wine, milk, honey, and flowers. **prīscō mōre parentum** by the ancient custom of our forefathers. **trādita sunt** are handed over, given. **trīstī mūnere** as a sad duty (**mūnus** means gift; duty: both meanings are implied here). **ad īnferiās** to (be) a sacrifice to the dead.
9	**multum mānantia** dripping much = all wet.
10	**in perpetuum** for ever.

Troy

1 l.4 **nēquīquam**: why 'in vain'? **cinerem**: why 'ashes'?
2 l.10 **in perpetuum, frāter, avē atque valē**: explain what Catullus means.
3 How successful do you consider this poem? What feelings does it convey? Does it strike the right note for the occasion? Does it say too much, or too little, or just enough?
4 Compare the language Catullus used in this poem with that of Poem 10 (**Vārus mē meus ad suōs amōrēs**), pp. 120–2. What differences do you see? Explain why Catullus uses such different sorts of language in these two poems.
5 Which of the poems of Catullus you have read do you like best? Explain your preference.
6 Discuss the range of subjects on which Catullus writes and the different styles of writing he uses for different subjects.

Exercise 11.5

Translate into Latin
1 Catullus was persuaded to leave Rome and go to Bithynia.
2 When he arrived there, he made many friends.
3 He had to stay a whole year but he left as soon as he was allowed.
4 He would have stayed longer, if the place had pleased him.
5 He told his friends that he was going to sail to the famous cities of Asia.

Exercise 11.6

Translate into Latin
(Catullus writes from Bithynia to Veranius, who is serving in Asia.)

Dear Veranius,
            How are you? Does Asia please you? I don't think much of Bithynia. We have a very bad governor, who does nothing to help his staff. As soon as I have finished my year's duty, I shall leave and sail to Asia. I hope to meet you and Fabullus there. I have bought a smashing yacht. If you were to come with me, we could sail to all the famous cities. At the beginning of spring I shall write again to tell you when I am going to set out. Look after yourself.
    Yours, Catullus.

I think much of = I value highly	governor **praetor, praetōris**
year's duty **annuum mūnus**	yacht **phasēlus, -ī**, *m.*

127

CHAPTER XII

Aeneas arrives at Carthage

*The Greeks had laid siege to Troy for ten years before they finally
captured it by trickery. On the dreadful night when the city was sacked,
the ghost of Hector, who had been Troy's greatest warrior, appeared to
his cousin Aeneas in a dream and told him to flee from the burning
ruins. He must take with him the sacred emblems and the guardian gods
of Troy and found a home for them in a new city across the sea.*

*Aeneas obeys and wanders with his followers for seven years. In
all this time, he fails to reach the place where Fate and the gods wish
him to build his city. Only when he does so will he fulfil his mission by
founding a new Troy from which Rome will spring.*

*Virgil begins his story about Aeneas as his fleet is at last
approaching the destined land of Italy.*

Juno, queen of the gods, loves Carthage, but hates the Trojans.

    urbs antīqua fuit (Tyriī tenuēre colōnī)
    Karthāgō, Italiam contrā Tiberīnaque longē
    ōstia, dīves opum studiīsque asperrima bellī,
    quam Iūnō fertur terrīs magis omnibus ūnam
5   posthabitā coluisse Samō: hīc illius arma,
    hīc currus fuit; hoc rēgnum dea gentibus esse,
    sī quā fāta sinant, iam tum tenditque fovetque.
    prōgeniem sed enim Trōiānō ā sanguine dūcī
    audierat Tyriās ōlim quae verteret arcēs;
10   hinc populum lātē rēgem bellōque superbum
    ventūrum excidiō Libyae; sīc volvere Parcās.
    hīs accēnsa super iactātōs aequore tōtō

Virgil

The bay of Carthage

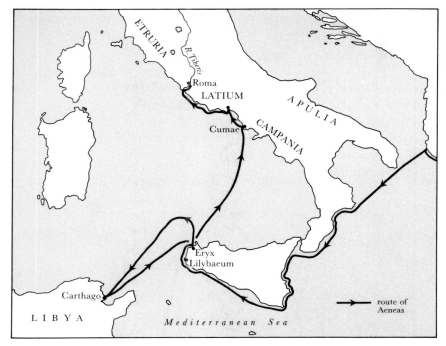

1 **Tyriī tenuēre (= tenuērunt) colōnī** colonists from Tyre held it, i.e. lived there. The Carthaginians are repeatedly called Tyrians because they come from Tyre, the Phoenician city on the coast of Syria.

2–3 *contrā Italiam Tiberīnaque ōstia longē* facing Italy and the mouth of the Tiber (i.e. Italy where the Tiber runs into the sea) at a distance – see map. **dīves opum** rich in resources.

4–5 **quam Iūnō fertur . . . ūnam . . . coluisse** which Juno is said to have loved more than any other land. **posthabitā Samō** with (even) Samos coming second. On Samos, an Aegean island, there was a famous temple to Juno.

5–6 **hīc illius arma,/hīc currus fuit**: i.e. in the temple was a statue of Juno armed and in a chariot.

6–7 *iam tum dea tendit fovetque* (intends and nurtures) *hoc esse rēgnum* (this to be a kingdom = to be the ruler) *gentibus*. **sī quā** if in any way.

8 **sed enim** but in fact (take at start of line).

 prōgeniem . . . Trōiānō ā sanguine dūcī offspring was being drawn from Trojan blood = a race was even now springing from Trojan blood.

9 *quae ōlim verteret Tyriās arcēs* which would one day overturn the Tyrians' citadels. The reference is to the Punic Wars.

10–11 indirect statement continues after **audierat**. **lātē rēgem** wide-ruling.

11 **ventūrum (esse)**. **excidiō** for a destruction to. . . (predicative dative) = to destroy. **sīc volvere Parcās** thus the Fates ordained. **ventūrum (esse)** and **volvere** are infinitives in indirect statement, part of what Juno had heard.

12 **hīs accēnsa super** inflamed (= enraged) over these things.

Trōas, rēliquiās Danaum atque immītis Achillī,
arcēbat longē Latiō, multōsque per annōs
15 errābant āctī fātīs maria omnia circum.
tantae mōlis erat Rōmānam condere gentem.

Juno persuades Aeolus, the god of the winds, to stir up a terrible storm.
The Trojan fleet, which has set sail from Sicily and has now almost
reached its destination of Italy, is driven off course and suffers much
damage and loss. Neptune calms the storm and Aeneas and most of his
men reach the coast of Libya safely. They are utterly demoralized but
Aeneas puts a brave face on things.

The storm

Iūnō

Jupiter now sends his messenger Mercury down to Libya to make
sure that the Carthaginians give Aeneas a friendly welcome. Venus,
disguising herself as a huntress, meets her son Aeneas as he explores the
country he has come to. She tells him the history of Dido, queen of
Carthage.

'Pūnica rēgna vidēs, Tyriōs et Agēnoris urbem;
sed fīnēs Libycī, genus intractābile bellō.
imperium Dīdō Tyriā regit urbe profecta,
20 germānum fugiēns. longa est iniūria, longae
ambāgēs; sed summa sequar fastīgia rērum.
huic coniūnx Sychaeus erat, dītissimus agrī
Phoenīcum, et magnō miserae dīlēctus amōre,
cui pater intāctam dederat prīmīsque iugārat
25 ōminibus. sed rēgna Tyrī germānus habēbat
Pygmaliōn, scelere ante aliōs immānior omnīs.

130

13	**Trōas** the Trojans (accusative plural). **rēliquiās Danaum atque immītis Achillī** = the remnants left by the Greeks and savage Achilles. **Danaī** = the Greeks. **Achillī:** genitive.
14	**Latiō:** Latium is the area South of the Tiber; see map.
15	*circum omnia maria.*
16	**tantae mōlis erat** it was a matter involving (literally, of) such effort . . . .

Tyre

17	**Pūnica** Carthaginian. **Agēnoris urbem**: Carthage (Agenor was an ancestor of Dido).
18	**fīnēs (sunt) Libycī** the surrounding country is Libyan, i.e. belonging to native Africans. The colonizing Carthaginians are hedged in by enemies. **intractābile bellō** fierce in war.
20–1	**germānum** (her) brother. **longa** (a) long (story of). **longae ambāgēs** a long involved story. **summa sequar fastīgia rērum** I shall follow (= trace) the main outlines of the story.
22–3	**dītissimus agrī/Phoenīcum** the wealthiest of the Phoenicians (i.e. Tyrians) in land. **magnō miserae dīlectus amōre** loved by great love of the unhappy girl = greatly loved by the unhappy girl.
24–5	**intāctam:** untouched = as a virgin. **prīmīs iugārat (=iugāverat) ōminibus** had joined in her first wedding ceremony (i.e. she had not been married before).
25	**germānus** (her) brother. **rēgna Tyrī** the kingdom of Tyre (**rēgna**: poetic plural for singular).
26	**immānior** more dreadful. **omnīs = omnēs** (-īs is a common variant for -ēs in the acc. plur. of the 3rd decl.).

quōs inter medius vēnit furor. ille Sychaeum
impius ante ārās atque aurī caecus amōre
clam ferrō incautum superat, sēcūrus amōrum
30 germānae; factumque diū cēlāvit et aegram
multa malus simulāns vānā spē lūsit amantem.
ipsa sed in somnīs inhumātī vēnit imāgō
coniugis ōra modīs attollēns pallida mīrīs;
crūdēlīs ārās trāiectaque pectora ferrō
35 nūdāvit, caecumque domūs scelus omne retēxit.
tum celerāre fugam patriāque excēdere suādet
auxiliumque viae veterēs tellūre reclūdit
thēsaurōs, ignōtum argentī pondus et aurī.
hīs commōta fugam Dīdō sociōsque parābat.
40 conveniunt quibus aut odium crūdēle tyrannī
aut metus ācer erat; nāvīs, quae forte parātae,
corripiunt onerantque aurō. portantur avārī
Pygmaliōnis opēs pelagō; dux fēmina factī.
dēvēnēre locōs ubi nunc ingentia cernēs
45 moenia surgentemque novae Karthāginis arcem.'

*The disguised Venus now tells Aeneas that all the men and ships he had
thought lost have in fact escaped the storm. She sends him on his way.*

corripuēre viam intereā, quā sēmita mōnstrat.
iamque ascendēbant collem, quī plūrimus urbī
imminet adversāsque aspectat dēsuper arcēs.
mīratur mōlem Aenēās, māgālia quondam,
50 mīrātur portās strepitumque et strāta viārum.
īnstant ārdentēs Tyriī: pars dūcere mūrōs
mōlīrīque arcem et manibus subvolvere saxa,
pars optāre locum tēctō et conclūdere sulcō;
iūra magistrātūsque legunt sānctumque senātum.
55 hīc portūs aliī effodiunt; hīc alta theātrī
fundāmenta locant aliī, immānīsque columnās
rūpibus excīdunt, scaenīs decora alta futūrīs.
quālis apēs aestāte novā per flōrea rūra
exercet sub sōle labor, cum gentis adultōs
60 ēdūcunt fētūs, aut cum līquentia mella
stīpant et dulcī distendunt nectare cellās,
aut onera accipiunt venientum, aut agmine factō
īgnāvum fūcōs pecus ā praesēpibus arcent:
fervet opus redolentque thymō fragrantia mella.
65 'ō fortūnātī, quōrum iam moenia surgunt!'
Aenēās ait et fastīgia suspicit urbis.

mīrātur mōlem Aenēās

apis

132

27	**quōs inter = inter quōs:** referring to Pygmalion and Sychaeus.
29–30	**incautum** off his guard. **sēcūrus amōrum/germānae** caring nothing for his sister's love. **aegram** sick = distraught.
31	**malus** wickedly, in his wickedness. **vānā spē lūsit amantem** he deluded the loving (wife) with vain hope.
32–3	**ipsa . . . inhumātī . . . imāgō/coniugis** the ghost (itself) of her unburied husband.
33	**ōra modīs attollēns pallida mīrīs** lifting his face to her, pale in wondrous ways = strangely pale. **ōra:** poetic plural for singular.
34	**crūdēlīs = crūdēlēs. trāiecta** transfixed.
35	**nūdāvit** he stripped bare = he revealed. **caecum** hidden. **retēxit** uncovered.
36	**celerāre** speed up = embark hastily upon. **suādet** he urges (her to . . .)
37–8	**auxilium viae veterēs tellūre reclūdit/thēsaurōs** he revealed to her ancient treasures (hidden) in the ground (as a) help for the journey.
40	**conveniunt (eī) quibus. tryannī** of the tyrant, i.e. of Pygmalion.
41	**nāvīs = nāvēs. parātae (erant).**
42	**onerant** they load. **portantur** are carried off.
43	**pelagō** over the sea. *dux factī (fuit) fēmina.*
44	**dēvēnēre = dēvēnērunt** they came down (from the sea), = they landed at. **locōs = in locōs.**
46	**corripuēre (= corripuērunt) viam** they (i.e. Aeneas and his companion) hastened on their way. **quā** where. **sēmita** the pathway.
47	**plūrimus** with its huge mass.
48	**aspectat** looks towards. **adversās** = facing them.
49	**mōlem:** i.e. the massive structures. **māgālia quondam** once huts.
50	**strāta** the paving.
51–3	**īnstant** press on (with the work). **ārdentēs** burning, i.e. enthusiastically. **pars . . . pars** some . . . others . . . **dūcere** (= build), **mōlīrī** (= toil at), **subvolvere** (= roll up), **optāre, conclūdere** (= enclose) are all historic infinitives. **tēctō** for a building. **sulcō** with a trench. This was the way the Romans marked out the site for city walls.
54	**legunt** choose, elect. They establish laws and elect magistrates and a government.
55–6	**alta theātrī/fundāmenta locant** they lay the deep foundations of a theatre.
56–7	**immānīs (=immānēs) columnās/rūpibus excīdunt** they cut enormous columns from the rocks. **scaenīs decora alta futūrīs** lofty adornments for the stage which is going to be (built).
58–9	**quālis apēs . . . exercet . . . labor** like the work which keeps bees busy. **quālis** (= what sort of) commonly introduces a simile: **quālis labor** what sort of work = like the work which . . . . **flōrea** flowery.
59–60	**gentis adultōs . . . fētūs** the young offspring of their race.
60–1	**līquentia mella/stīpant** they cram in the liquid honey. **dulcī distendunt nectare cellās** they fill to bursting the honeycomb cells with sweet nectar.
62	**venientum = venientium** i.e. of the bees coming back from the meadows.
63	**īgnāvum fūcōs pecus ā praesēpibus arcent** they keep off the drones (**fūcōs**), an idle tribe, from their hives.
64	**fervet:** literally, boils. **redolent thymō fragrantia mella** the sweet-smelling honey is scented with thyme.
66	**ait** says. **fastīgia suspicit** looks down on the tops of the buildings.

(V)

āra, -ae, *f.*	altar	**fīnēs, fīnium,** *m.pl.*	territory, country
asper, aspera, asperum	rough, harsh, dangerous	**hinc**	from here
avārus-a-um	greedy, miserly	**impius-a-um**	impious, unholy
caecus-a-um	blind, hidden	**memor, memoris** + genitive	mindful of, recalling
cernō, cernere, crēvī, crētum	I perceive, decide	**mōnstrō** (1)	I show
coniūnx, coniugis, *m. & f.*	husband, wife	**spernō, spernere, sprēvī, sprētum**	I despise

(G) Word building

Make sure that you know the following compounds of

fugiō, fugere, fūgī, fugitum

effugiō	I escape			**profugiō**	I flee away
cōnfugiō	I flee, take refuge	**perfugiō**	I flee away, take refuge	**refugiō**	I retreat
diffugiō	I scatter				

(G) '-īs' for '-ēs'

In Latin poetry, the accusative plural of third declension nouns and adjectives often ends in **-īs** instead of **-ēs**, e.g. **omnīs** for **omnēs**, **crūdēlīs** for **crūdēlēs**.

(G) Relative with the subjunctive—consecutive and generic

You have met the relative with the subjunctive used to express purpose. The relative pronoun is also followed by the subjunctive when it means 'such that', 'of the sort that', e.g.

nōn is est quī tālia dīcat
He is not the sort of person to say (literally, who would say) such things.
sunt quī nimis bibant
There are people who drink too much, i.e. some people drink too much.
dignus erat quī pūnīrētur
He was worthy who should be punished, i.e. he deserved to be punished.
prūdentior est quam quī nimis bibat
He is more sensible than a man who would drink too much, i.e. he is too sensible to drink too much.

Exercise 12.1

Translate
1 Aenēās nōn is est quī perīculum vītet.
2 sunt quī impia faciant ut aurum possideant.
3 Dīdō digna erat quae rēgīna fieret.
4 Aenēās fortior erat quam quī dēspērāret.
5 sunt quī hospitēs cōmiter salūtent; Polyphēmus nōn est quī ita faciat.

134

*Aeneas proceeds into the city. He is greatly moved by the
fact that in this remote place the walls of the temple of
Juno have been decorated with scenes from the Trojan
War. As he gazes at these, Dido appears.*

Exercise 12.2

Read the following passage and then answer the questions below it

mediā testūdine templī

```
    haec dum Dardaniō Aenēae mīranda videntur,
    dum stupet obtūtūque haeret dēfīxus in ūnō,
    rēgīna ad templum, fōrmā pulcherrima Dīdō,
    incessit magnā iuvenum stīpante catervā.
5   quālis in Eurōtae rīpīs aut per iuga Cynthī
    exercet Dīāna chorōs, quam mīlle secūtae
    hinc atque hinc glomerantur Oreādes; illa pharetram
    fert umerō gradiēnsque deās superēminet omnīs
    (Lātōnae tacitum pertemptant gaudia pectus):
10  tālis erat Dīdō, tālem sē laeta ferēbat
    per mediōs īnstāns operī rēgnīsque futūrīs.
    tum foribus dīvae, mediā testūdine templī,
    saepta armīs soliōque altē subnīxa resēdit.
    iūra dabāt lēgēsque virīs, operumque labōrem
15  partibus aequābat iūstīs aut sorte trahēbat.
```

1   **Dardaniō Aenēae** to Trojan Aeneas. Dardanus was the founder of the
    Trojan race.
2   **stupet** is dumbfounded. **obtūtū haeret dēfīxus in ūnō** he sticks, held
    fast in a single gaze = stands there, rooted to the spot in total
    concentration.
4   **stīpante** pressing (round her). **catervā** company.
5   **Eurōtae** of the Eurotas (the river on which Sparta stood). **iuga Cynthī**
    the ridges of Cynthus (the hill on Delos). See Part II, p. 84.
6   **exercet Dīāna chorōs** Diana sets (her followers) singing and dancing.
7   **hinc atque hinc** on this side and on that. **glomerantur** gather
    together. **Oreādes,** *f.pl.* mountain nymphs. **pharetram** quiver.
8   **umerō** shoulder. **gradiēns** as she goes. **deās**: i.e. the mountain
    nymphs. **superēminet** overtops.
9   **pertemptant gaudia** joy thrills . . . **Lātōna**: the mother of Diana.
10 **tālis** picks up **quālis**, the word which began the simile in 1.5.
11 **īnstāns** (+ dative) pressing on with.
12 **foribus dīvae** at the doors of the goddess. **mediā testūdine** beneath the
    centre of the vault.
13 **saepta** hedged about. **soliō altē subnīxa** taking her place high up on
    her throne. **resēdit** she sat down.
15 **partibus aequābat iūstīs** she was making equal in fair parts = she was
    dividing up justly and equally. **sorte trahēbat** was assigning by lot.

1 Translate ll.1–4.
2 Describe what is going on in the simile (ll.5–9). What are we being
  told about Dido through this comparison to Diana?

3 What does Dido do (a) on the way to her throne, and (b) after she has sat there? In what ways do these activities strike you as characteristic of Dido as she has been described earlier in this chapter? What do you feel about Dido, both as a queen and as a woman?

4 What case and what number (i.e singular or plural) are **omnīs** (l.8), **tālis** (l.10) and **futūrīs** (l.11)?

5 Scan ll.10–11.

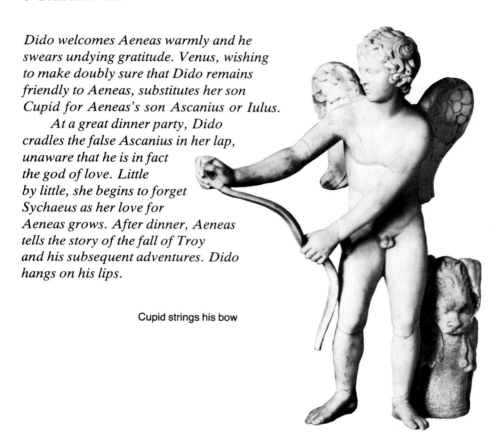

Dido welcomes Aeneas warmly and he swears undying gratitude. Venus, wishing to make doubly sure that Dido remains friendly to Aeneas, substitutes her son Cupid for Aeneas's son Ascanius or Iulus.

At a great dinner party, Dido cradles the false Ascanius in her lap, unaware that he is in fact the god of love. Little by little, she begins to forget Sychaeus as her love for Aeneas grows. After dinner, Aeneas tells the story of the fall of Troy and his subsequent adventures. Dido hangs on his lips.

Cupid strings his bow

Exercise 12.3

Translate into Latin

1 Aeneas asked his mother where he was.

2 She replied that he was in the territory of queen Dido.

3 If you go to the city, you will meet the queen.

4 You are fortunate because you are building a city.

5 I must find a land where I myself shall found a new city.

Dido and Aeneas at
the banquet

Exercise 12.4

Translate into Latin

Pygmalion, the brother of Dido, killed her husband Sychaeus in order
that he might seize his gold. He concealed the deed so cleverly
(**callidē**) that his sister did not find out about it. But she was very sad
because she loved her husband greatly and thought that he had gone
away. One (i.e. on a certain) night, the ghost of Sychaeus appeared to
her. He told her what had happened and advised her to flee. She
obeyed him and showed herself a most outstanding leader.

ipsa sed in somnīs inhumātī vēnit imāgō

CHAPTER XIII

Dido and Aeneas

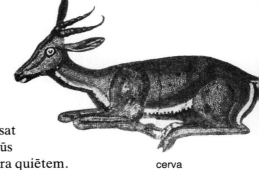

Dido is wounded by her love.
at rēgīna gravī iamdūdum saucia cūrā
vulnus alit vēnīs et caecō carpitur ignī.
multa virī virtūs animō multusque recursat
gentis honōs: haerent īnfīxī pectore vultūs
5   verbaque, nec placidam membrīs dat cūra quiētem.

cerva

*Encouraged by her sister Anna, Dido gives free rein to her
passion for Aeneas. She tries vainly to win the gods' favour by
sacrifices.*
heu, vātum īgnārae mentēs! quid vōta furentem,
quid dēlūbra iuvant? ēst mollīs flamma medullās
intereā et tacitum vīvit sub pectore vulnus.
ūritur īnfēlīx Dīdō tōtāque vagātur
10   urbe furēns, quālis coniectā cerva sagittā,
quam procul incautam nemora inter Crēsia fīxit
pāstor agēns tēlīs līquitque volātile ferrum
nescius: illa fugā silvās saltūsque peragrat
Dictaeōs; haeret laterī lētālis harundō.
15   nunc media Aenēān sēcum per moenia dūcit
Sīdoniāsque ostentat opēs urbemque parātam,
incipit effārī mediāque in vōce resistit;
nunc eadem lābente diē convīvia quaerit
Iliacōsque iterum dēmēns audīre labōrēs
20   exposcit pendetque iterum nārrantis ab ōre.
post ubi dīgressī, lūmenque obscūra vicissim
lūna premit suādentque cadentia sīdera somnōs,
sōla domō maeret vacuā strātīsque relictīs
incubat. illum absēns absentem auditque videtque,
25   aut gremiō Ascanium genitōris imāgine captā
dētinet, īnfandum sī fallere possit amōrem.
nōn coeptae adsurgunt turrēs, nōn arma iuventūs
exercet portūsve aut prōpugnācula bellō
tūta parant: pendent opera interrupta minaeque
30   mūrōrum ingentēs aequātaque māchina caelō.

pāstor nescius

1	**iamdūdum** now for a long time. **saucia** wounded. **cūrā**: often used of the care of love.
2	**alit vēnīs** feeds with her life-blood. **carpitur** is wasted, pines away.
3–4	**multa virī virtūs** many (a thought of) the courage of the man. **recursat** comes swiftly back. **multus gentis honōs** many (a thought of) the great distinction of his race (**honōs = honor**) **haerent īnfīxī vultūs** his features (**vultūs**) remain firmly fixed in . . . .
5	**placidam** gentle.
6	**vātum** of the soothsayers (who are advising her how to win the gods' favour). **quid vōta (iuvant)**. **furentem** her in her frenzy.
7	**dēlūbra** shrines. **ēst** eats. **mollīs . . . medullās** the soft marrow of her bones.
8	**tacitum** she cannot speak of her love to Aeneas.
10	**coniectā . . . sagittā** when the arrow has been fired.
11	**incautam** off her guard. **nemora inter Crēsia** amid the woods of Crete. **fīxit** has transfixed.
12	**agēns** pursuing (her). **volātile** winged.
13–14	**silvās saltūsque . . . Dictaeōs** the wooded defiles (literally, the woods and defiles) of Crete. Dicte is a mountain on Crete. **lētālis harundō** the deadly shaft (literally: reed).
15	**Aenēān**: accusative.
16	**Sīdoniās**: of Tyre. Virgil refers to Tyre and Sidon as if they were the same place. **ostentat** shows off.
17	**effārī** to speak out, i.e. to declare her love. **resistit** stops.
18	**eadem . . . convīvia** i.e. repetitions of the first banquet at which Aeneas had told his tale.
19	**Iliacōs** Trojan. **dēmēns** in her madness.
21–2	**post** later. **dīgressī (sunt)** they have parted. **vicissim** in turn. **suādent** urge.
23–4	**maeret** she mourns. **strātīs relictīs/incubat** she lies on the couch he has left (i.e. had reclined on at the banquet).
25	**gremiō** in her lap. **genitōris** of his father.
26	**dētinet** she clings on to. **sī** if = in the hope that. **īnfandum . . . amōrem** a love she could not speak of. She tries to hoodwink love and herself.
27	**coeptae** which had been begun.
28–9	**prōpugnācula bellō/tūta** the ramparts bringing safety in war. **parant**: subject *they*, the young men.
29–30	**pendent** are suspended = are idle. **interrupta** broken off. **mīnae mūrōrum ingentēs**: the vast threats of the walls = the vast threatening walls'. **aequāta māchina caelō** the crane towering to the sky.

The hunt

*Juno plots with Venus to strengthen the union between Dido and
Aeneas. Her aim is to keep Aeneas in Carthage and thus prevent him
from fulfilling his destiny in Italy. Venus is glad to see the bonds of
hospitality strengthened.*

> *Dido and Aeneas go hunting.*
> Oceanum intereā surgēns Aurōra relīquit.
> it portīs iubare exortō dēlēcta iuventūs,
> rētia rāra, plagae, lātō vēnābula ferrō,
> Massȳlīque ruunt equitēs et odōra canum vīs.

35 rēgīnam thalamō cunctantem ad līmina prīmī
Poenōrum exspectant, ostrōque īnsignis et aurō
stat sonipēs ac frēna ferōx spūmantia mandit.
tandem prōgreditur magnā stīpante catervā;
cui pharetra ex aurō, crīnēs nōdantur in aurum,

40 aurea pupuream subnectit fībula vestem.
nec nōn et Phrygiī comitēs et laetus Iūlus
incēdunt. ipse ante aliōs pulcherrimus omnīs
īnfert sē socium Aenēās atque agmina iungit.
quālis ubi hībernam Lyciam Xanthīque fluenta

45 dēserit ac Dēlum māternam invīsit Apollō
īnstauratque chorōs, mixtīque altāria circum
Crētēsque Dryopēsque fremunt pictīque Agathyrsī:
ipse iugīs Cynthī graditur mollīque fluentem
fronde premit crīnem fingēns atque implicat aurō,

50 tēla sonant umerīs: haud illō sēgnior ībat
Aenēās, tantum ēgregiō decus ēnitet ōre.
postquam altōs ventum in montīs atque invia lustra,
ecce ferae saxī dēiectae vertice caprae
dēcurrēre iugīs; aliā de parte patentīs

55 trānsmittunt cursū campōs atque agmina cervī
pulverulenta fugā glomerant montīsque relinquunt.
at puer Ascanius mediīs in vallibus ācrī
gaudet equō iamque hōs cursū, iam praeterit illōs,
spūmantemque darī pecora inter inertia vōtīs

60 optat aprum, aut fulvum dēscendere monte leōnem.

Apollo

Ascanius spūmantem
darī optat aptum

31	**Oceanum**: the stream of Ocean which surrounded the world. **Aurōra**: the goddess of dawn.

31   **Oceanum**: the stream of Ocean which surrounded the world.   **Aurōra**: the goddess of dawn.

32   **iubare exortō** when the sun's beams arose.

33   **rētia rāra** wide-meshed nets.   **plagae** trap-nets.   **lātō vēnābula ferrō** broad-bladed hunting-spears. The horsemen pour along with their gear of nets, spears, etc.

34   **Massȳlī** African. The Massyli were a North African tribe.   **odōra canum vīs** the keen-scented strength of dogs = keen-scented, strong dogs.

35   **thalamō** in her bedroom.

36   **ostrō īnsignis** standing out in purple . . . .

37   **sonipēs** (her) steed.   **frēna ferōx spūmantia mandit** proudly champs the foaming bit.

39   **cui pharetra (est)** to whom there is a quiver = she has a quiver.   **crīnēs nōdantur in aurum** her hair is knotted onto a golden clasp.

40   **aurea purpuream subnectit fībula vestem** a golden brooch fastens her purple dress.

41   **Phrygiī** Trojan.

43   **īnfert sē socium** brings himself as a companion = goes to join her.

44   **hībernam Lyciam** Lycia, his winter home.   **Xanthī fluenta** the streams of Xanthus (in Lycia, where Apollo dwelt at Patara in the winter).

45   **dēserit** leaves.   **Dēlum māternam** his mother's Delos = Delos, where his mother gave him birth.

46   **īnstaurat chorōs** starts up the dance afresh.   **mixtī altāria circum** mingled around the altars.

47   **fremunt** make a noise.   **pictī** tattooed. The Dryopes were from Northern Greece, the Agathyrsi from Scythia; Cretans, Dryopes and Agathyrsi all gather at Delos for Apollo's festival.

48   **ipse**: Apollo himself, the master.

48–9 **mollī fluentem/fronde premit crīnem fingēns atque implicat aurō** shaping his flowing hair, he garlands (literally, presses) it with soft foliage and entwines it with gold, i.e. he puts on his hair (**crīnem**) a crown of leaves entwined with gold wire.

50   **umerīs** on his shoulders.   **haud sēgnior** no more sluggish than . . . = even more briskly than . . . .

51   **ēnitet** shines forth.

52   **ventum (est)**.   **invia lustra** the trackless lairs (of the beasts).

53   **ecce** look!   **ferae . . . caprae** wild goats.   **dēiectae vertice** dislodged from the top of . . . The beaters have driven the goats down.

54–5 **dēcurrēre** = **dēcurrērunt**.   **trānsmittunt** they cross.   **cervī** stags (subject of **trānsmittunt** and **glomerant**).

56   **pulverulenta** dusty.   **glomerant** mass (transitive).

57   Aeneas's son is known both as Ascanius and Iulus.

58   **hōs . . . illōs**: referring to his fellow hunters.

59–60 **spūmantem . . . aprum** a foaming wild boar.   **pecora inter inertia** amid (these) harmless creatures. He prays to be given a serious challenge.   **fulvum** tawny.

intereā magnō miscērī murmure caelum
incipit. īnsequitur commixtā grandine nimbus,
et Tyriī comitēs passim et Trōiāna iuventūs
Dardaniusque nepōs Veneris dīversa per agrōs
65 tēcta metū petiēre; ruunt de montibus amnēs.
spēluncam Dīdō dux et Trōiānus eandem
dēveniunt. prīma et Tellūs et prōnuba Iūnō
dant signum; fulsēre ignēs et cōnscius aethēr
cōnubiīs, summōque ululārunt vertice Nymphae.
70 ille diēs prīmus lētī prīmusque malōrum
causa fuit; neque enim speciē famāve movētur
nec iam fūrtīvum Dīdō meditātur amōrem:
coniugium vocat, hōc praetexit nōmine culpam.

spēluncam eandem dēveniunt

*The malevolent goddess Rumour brings the news of the love affair
to an African king, the Gaetulian Iarbas. He has himself been
rejected as a suitor by Dido and is especially indignant that she
preferred to himself a man he thinks of as an effeminate oriental.
He prays to his father Jupiter to take action.*

    *Jupiter is angry to discover that Aeneas is neglecting the will
of Destiny and he sends Mercury down to him with a stinging
rebuke. Mercury flies to Carthage.*

61	**miscērī** to be disturbed. **murmure** with rumbling.
62	**commixtā grandine** with hail mixed in = mixed with hail. **nimbus** rain-cloud.
63	**passim** in every direction.
64–5	**Dardanius nepōs Veneris** the Trojan grandson of Venus, i.e. Ascanius. **dīversa . . . tēcta** various shelters. **petiēre** = **petiērunt.** **amnēs** rivers.
66	**spēluncam** cave: supply **ad.** *Dīdō et dux Trōiānus.*
67	**prīma Tellūs** primeval Earth. **prōnuba** goddess of marriage.
68–9	**fulsēre** = **fulsērunt** blazed. **cōnscius aethēr/cōnubiīs** the upper air, the witness to the marriage. **summō ululārunt (= ululāvērunt) vertice Nymphae** the Nymphs cried out on the mountain top. Virgil conjures up the language and ritual of a Roman wedding. Earth and Juno give the signal for the ceremony to begin. The flashes of lightning are the marriage torches, the upper air is the witness, and the Nymphs sing the marriage song.
70	**lētī** of death.
71	**speciē** by appearance = by how things appear.
72	**nec iam fūrtīvum . . . meditātur amōrem** and no longer practises a secret love = and she no longer conceals her love.
73	**coniugium vocat** she calls it (i.e. her liaison with Aeneas) a marriage. **praetexit . . . culpam** she masks her sin.

A Roman wedding

ut prīmum ālātīs tetigit māgālia plantīs,
75 Aenēān fundantem arcēs ac tēcta novantem
cōnspicit. atque illī stēllātus iaspide fulvā
ēnsis erat Tyriōque ārdēbat mūrice laena
dēmissa ex umerīs, dīves quae mūnera Dīdō
fēcerat, et tenuī tēlās discrēverat aurō.
80 continuō invādit: 'tū nunc Karthāginis altae
fundāmenta locās pulchramque uxōrius urbem
exstruis? heu, rēgnī rērumque oblīte tuārum!
ipse deum tibi mē clārō dēmittit Olympō
rēgnātor, caelum ac terrās quī nūmine torquet:
85 ipse haec ferre iubet celerīs mandāta per aurās:
quid struis? aut quā spē Libycīs teris ōtia terrīs?
sī tē nūlla movent tantārum glōria rērum,
Ascanium surgentem et spēs hērēdis Iūlī
respice, cui rēgnum Italiae Rōmānaque tellūs
90 dēbētur.' tālī Cyllēnius ōre locūtus
mortālīs vīsūs mediō sermōne relīquit
et procul in tenuem ex oculīs ēvānuit auram.
    at vērō Aenēās aspectū obmūtuit āmēns,
arrēctaeque horrōre comae et vōx faucibus haesit.
95 ārdet abīre fugā dulcīsque relinquere terrās,
attonitus tantō monitū imperiōque deōrum.
heu, quid agat? quō nunc rēgīnam ambīre furentem
audeat adfātū? quae prīma exordia sūmat?
atque animum nunc hūc celerem nunc dīvidit illūc
100 in partīsque rapit variās perque omnia versat.

Mercury

(V) absēns, absentis	absent	**sīdus, sīderis**, *n.*	constellation, star;
adhūc	still		time of year
coniugium, coniugiī, *n.*	marriage	**speciēs, speciēī**, *f.*	appearance
cunctor (1)	I delay	**ūrō, ūrere, ussī,**	I burn (transitive)
exerceō (2)	I keep busy, exercise	**ustum**	
haereō, -ēre, haesī,	I stick		
haesum			
heu!	alas!		
iugum, -ī, *n.*	ridge, hill; yoke		
iuventūs, iuventūtīs, *f.*	youth		
leō, leōnis, *m.*	lion		
omnīnō	altogether, completely		
pāstor, pāstōris, *m.*	shepherd		
propter + accusative	on account of		

74	**ālātīs ... plantīs** with his winged feet.
75	**fundantem** laying foundations for. **tēcta novantem** renewing buildings = building new dwellings.
76-7	**stēllātus iaspide fulvā/ēnsis** a sword starred with tawny jasper.
77	**Tyriō ārdēbat mūrice laena** (his) cloak was glowing with Tyrian dye.
78-9	**ex umerīs** from his shoulders. *mūnera quae dīves Dīdō fēcerat.* **tenuī tēlās discrēverat aurō** had interwoven the fabric with fine (strands of) gold.
80	**invādit:** Mercury 'went for' him.
81	**fundāmenta locās** are you laying the foundations? **uxōrius** enslaved to a wife.
82	**extruis** are you building? **heu, ... oblīte** alas, you who have forgotten ... = alas that you have forgotten ....
83-4	**ipse deum ... rēgnātor** the ruler of the gods himself.
84	**caelum ... quī ... torquet** = **quī caelum ... torquet. nūmine torquet** controls with his divine power.
85	**iubet (mē).**
86	**struis** are you aiming at? **teris ōtia** are you wasting your time?
88	**hērēdis** of (your) heir.
89	**respice** consider.
90	**dēbētur** is owed, i.e. rule over Italy and the soil of Rome are being held in trust for Ascanius. Aeneas must pay his debt. **tālī ... ōre** with such a mouth = with such words. **Cyllēnius:** Mercury, born on Mount Cyllene in Arcadia.
91	**mortālīs vīsūs** the sight of men.
92	**in tenuem ... ēvānuit auram** he vanished into thin air.
93	**aspectū obmūtuit āmēns** was beside himself (**āmēns**) and dumbfounded at what he had seen.
94	**arrēctae (sunt) horrōre comae** his hair stood on end with dread. **faucibus haesit** stuck in his throat.
95	**ārdet** he is on fire, he longs.
96	**attonitus** thunderstruck, dazed.
97-8	**quid agat?** what was he to do? **quō ... ambīre ... audeat adfātū** with what words (**quō adfātū**) could he have the courage to approach ... **prīma exordia** beginning (of a speech).
99	**hūc ... illūc** this way and that. **celerem** rapidly (adjective used adverbially).
100	**in partīs ... variās** to all sorts of considerations. **versat** turns.

G Word building

Make sure that you know the following compounds of
dūcō, dūcere, dūxī, ductum

addūcō	I bring to, induce	**indūcō**	I lead on, induce
condūcō	I collect, hire	**intrōdūcō**	I introduce, bring in
dēdūcō	I lead down	**prōdūcō**	I bring forward, prolong
dīdūcō	I separate	**subdūcō**	I remove, withdraw

Exercise 13.1

Translate

1 Dīdō Aenēān ōrat ut labōrēs ēius
   rūrsus sibi nārret.
2 Poenī Trōiānīquē māne profectī sunt
   ut vēnārentur.
3 Iuppiter valdē īrātus est; timet nē
   Aenēās Rōmae oblīvīscātur.

4 Libyam relinquāmus, ō sociī, et
   Italiam quaerāmus!
5 quōmodō Aenēās rēgīnam amantem
   dē imperiō Iovis certiōrem facere
   audēret?
6 sī mē relinquās, moriar.

Exercise 13.2

Read the following passage and answer the questions beneath it

*Aeneas makes preparations to go, but still cannot find the right way to
break the news to Dido. She senses what is happening, and her instinct
is confirmed by Rumour. In a frenzy, she summons Aeneas and speaks
these words to him:*

'dissimulāre etiam spērāstī, perfide, tantum
posse nefās tacitusque meā dēcēdere terrā?
nec tē noster amor nec tē data dextera quondam
nec moritūra tenet crūdēlī fūnere Dīdō?
5 quīn etiam hībernō mōlīris sīdere classem
et mediīs properās Aquilōnibus īre per altum,
crūdēlis? quid, sī nōn arva aliēna domōsque
ignōtās peterēs, et Trōia antīqua manēret,
Trōia per undōsum peterētur classibus aequor?
10 mēne fugis? per ego hās lacrimās dextramque tuam tē
(quandō aliud mihi iam miserae nihil ipsa relīquī),
per cōnūbia nostra, per inceptōs hymenaeōs,
sī bene quid dē tē meruî, fuit aut tibi quicquam
dulce meum, miserēre domūs lābentis, et istam,
15 ōrō, sī quis adhūc precibus locus, exue mentem.

Because of you, the Libyan tribes and the Nomad chieftains
Hate me, the Tyrians are hostile: because of you I have lost
My old reputation for faithfulness – the one thing that could have made me
Immortal. Oh, I am dying! To what, my guest, are you leaving me?
20 'Guest' – that is all I may call you now, who have called you husband.
Why do I linger here? Shall I wait till my brother, Pygmalion,
Destroys this place, or Iarbas leads me away captive?

(C. Day Lewis)

saltem sī qua mihī dē tē suscepta fuisset
ante fugam subolēs, sī quis mihi parvulus aulā
25 lūderet Aenēās, quī tē tamen ōre referret,
nōn equidem omnīnō capta ac dēserta vidērer.'

1   **dissimulāre** to cover up.   **spērāstī** = **spērāvistī**.   **perfide** traitor.

2   **nefās** a sinful thing.

3   *dextera (manus) quondam data* your right hand once given me. Bride
and groom joined hands at the marriage ceremony.

4   **crūdēlī fūnere** by a cruel death.

5   **quīn etiam** what is more, are you even . . .?   **hībernō . . . sīdere** in the
wintry season.   **mōlīris . . . classem** are you preparing your fleet?

6   **mediīs Aquilōnibus** amid the North winds of winter.   **altum** the high
sea.

7   **quid, sī** what if . . . = tell me, suppose that . . . .   **arva aliēna** foreign
fields.

8   **ignōtās** unknown.

9   **per undōsum . . . aequor** over the swollen sea. Dido implies that
Aeneas would not even return home (if home still existed) at this
season. Can he really be thinking of sailing into the unknown in
winter?

10   *per hās lacrimās . . . ego tē . . . ōrō* (1.15).

11   **quandō** since.

12   **cōnūbia** union.   **inceptōs hymenaeōs** the marriage we have begun.

13–14 **fuit aut tibi quicquam/dulce meum** or if there was anything about me
that was sweet to you.
miserēre pity (imperative of **misereor** + genitive).

14–15 **domūs lābentis** slipping house = doomed house   **istam . . . exue
mentem** put off that mind = give up that idea.

23–4 **saltem** at least.   **sī qua mihī dē tē suscepta fuisset/ante fugam subolēs** if
a child (**subolēs**) from you had been taken up by me (i.e. in my arms)
before your flight.

24   **parvulus** darling little.   **aulā** in my hall.

25   **tē . . . ōre referret** could bring you back with his face = could remind
me of you by his features.   **tamen**: i.e. though you yourself have
deserted me.

26   **equidem** I indeed.   **capta et deserta** cheated and abandoned.

1 How many times does the letter 's' appear in ll.1–2? What emotion is Virgil aiming to convey here?
2 What do you think Dido means when she uses the word **moritūra** in l.4? (See also l.19.)
3 Translate ll.5–9.
4 List the various ways in which she appeals to Aeneas in ll.3–4 and ll.10–17 (to 'hostile').
5 What conclusion does she draw from the fact that he is determined to leave despite the appalling conditions described in ll.5–9?
6 There are several changes of tone in this passage: sometimes Dido is furiously angry with Aeneas, sometimes she appeals to his sense of pity, etc. Try to trace and explain the changes. Quote from the Latin to illustrate your answer.
7 If you had been Aeneas, what would you have said in answer to Dido?

Exercise 13.3

Translate into Latin

1 She thinks that he is her husband, but she is wrong.
2 Go to Italy and build a new Troy there.
3 'What am I to do?' Aeneas asked himself. He was not able to decide.
4 Surely he will not leave the queen whom he loves so much?
5 If he is unwilling to leave Dido, Jupiter will punish him.

Exercise 13.4

Translate

Because the goddess Juno wished to stop Aeneas (accusative **Aenēan**), she persuaded Aeolus, the god of the winds, to send forth a huge storm in order to destroy the Trojan leader's fleet. When the storm arose, Aeneas and his men were very much afraid; but the god of the sea, Neptunus, angered by the insolence (**īnsolentia, -ae**, *f.*) of the winds who had invaded his territory, ordered them to return to the mountain in which they had been shut before. Thus Aeneas and his men were able to reach land and disembarked on the shores of Africa.

The story of Dido and Aeneas is told in a mosaic from Low Ham in Somerset

CHAPTER XIV

The death of Dido

Aeneas replies to Dido.

     ille Iovis monitīs immōta tenēbat
lūmina et obnīxus cūram sub corde premēbat.
tandem pauca refert: 'ego tē, quae plūrima fandō
ēnumerāre valēs, numquam, rēgīna, negābō
5   prōmeritam, nec mē meminisse pigēbit Elissae
dum memor ipse meī, dum spīritus hōs regit artūs.
prō rē pauca loquar. neque ego hanc abscondere fūrtō
spērāvī (nē finge) fugam, nec coniugis umquam
praetendī taedās aut haec in foedera vēnī.
10   mē sī fāta meīs paterentur dūcere vītam
auspiciīs et sponte meā compōnere cūrās,

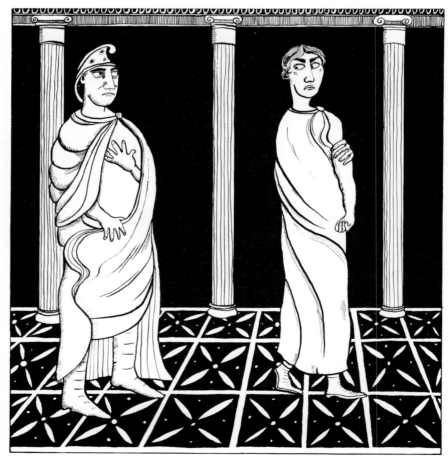

tandem pauca
refert

1–2 **monitīs** because of the advice. *lūmina immōta* his eyes motionless. **obnīxus** with great effort. **sub corde** deep in his heart. **premēbat** he stifled.

3–5 **refert** he replies. *ego numquam negābō tē prōmeritam (esse) plūrima quae fandō ēnumerāre valēs* I shall never deny that you have deserved (from me) as many things as you can possibly list in speaking. **mē meminisse pigēbit** shall I be sorry to remember. **Elissae:** Elissa is Dido's Tyrian name.

6 **dum (sum) memor.** **spīritus** breath.

7 **prō rē** for the situation = pleading my case. **abscondere fūrtō** to conceal with deceit'.

8 **nē finge** don't suppose (that).

8–9 **coniugis . . . praetendī taedās** did I hold out a bridegroom's torches. **haec . . . foedera** this bond, i.e. marriage. A torchlight procession escorted the bride and groom home after their wedding.

10–11 **meīs . . . auspiciīs** by my own authority = on my own terms. **sponte meā compōnere cūrās** to settle my cares to my liking.

'pius' Aeneas takes his father and son from the burning city of Troy (see l.32)

urbem Trōiānam prīmum dulcīsque meōrum
rēliquiās colerem, Priamī tēcta alta manērent,
et recidīva manū posuissem Pergama victīs.
15 sed nunc Italiam magnam Grȳnēus Apollō,
Italiam Lyciae iussēre capessere sortēs;
hic amor, haec patria est. si tē Karthāginis arcēs
Phoenissam Libycaeque aspectus dētinet urbis,
quae tandem Ausoniā Teucrōs cōnsīdere terrā
20 invidia est? et nōs fās extera quaerere rēgna.
mē patris Anchīsae, quotiēns ūmentibus umbrīs
nox operit terrās, quotiēns astra ignea surgunt,
admonet in somnīs et turbida terret imāgō;
mē puer Ascanius capitisque iniūria cārī,
25 quem rēgnō Hesperiae fraudō et fātālibus arvīs.
nunc etiam interpres dīvum Iove missus ab ipsō
(testor utrumque caput) celerīs mandāta per aurās
dētulit: ipse deum manifestō in lūmine vīdī
intrantem mūrōs vōcemque hīs auribus hausī.
30 dēsine mēque tuīs incendere tēque querēlīs;
Italiam nōn sponte sequor.'

*Dido replies furiously, with bitter complaints about his lack of
humanity and his ingratitude. She refuses to believe that the gods have
intervened and scornfully sends him on his way, hoping that he will die
at sea with her name on his lips. Her ghost will pursue him to the
underworld. She then runs from his sight.*

at pius Aenēās, quamquam lēnīre dolentem
sōlandō cupit et dictīs āvertere cūrās,
multa gemēns magnōque animum labefactus amōre,
35 iussa tamen dīvum exsequitur classemque revīsit.
tum vērō Teucrī incumbunt et lītore celsās
dēdūcunt tōtō nāvīs. natat uncta carīna,
frondentīsque ferunt rēmōs et rōbora silvīs
īnfabricāta fugae studiō.
40 migrantīs cernās tōtāque ex urbe ruentīs.
ac velut ingentem formīcae farris acervum
cum populant hiemis memorēs tēctōque repōnunt,
it nigrum campīs agmen praedamque per herbās
convectant calle angustō; pars grandia trūdunt
45 obnixae frūmenta umerīs, pars agmina cōgunt
castīgantque morās, opere omnis sēmita fervet.

formīca

12–13	**dulcīs meōrum/rēliquiās** the sweet remnants of my people. **colerem** I should be looking after. **tēcta,** *n.pl.* the palace.
14	**recidīva . . . Pergama** rebuilt Troy (**Pergama,** *n.pl.* Troy). **victīs** for the conquered.
15	**Grȳnēus Apollō** Grynian Apollo. At Grynium in Lydia there was a wood sacred to Apollo where Aeneas had consulted Apollo's oracle.
16	**Lyciae iussēre (=iussērunt) capessere sortēs** the oracle of Lycia has ordered me to make for . . . . Lycia was Apollo's winter home.
17–18	**tē . . . Phoenissam** you, a Tyrian. **Libycae aspectus dētinet urbis** the sight of the Libyan city (i.e. Carthage) holds you (here).
19–20	*quae tandem invidia est Teucrōs cōnsīdere Ausoniā terrā* why, I ask you (**tandem**), do you resent the Trojans' settling in Italian (**Ausoniā**) territory? (**quae . . . invidia est** literally: what is your grudge?). **et** too, also. **fās est** it is right. **extera** foreign.
21	**patris Anchīsae imāgō** (1.23) (**imāgō** the ghost of . . .). **quotiēns** every time that. **ūmentibus umbris** with moist shadows.
22	**operit** covers. **astra ignea** fiery stars.
23	**turbida** troubled.
24	**Ascanius** (the thought of) Ascanius. Supply 'admonet'. **capitis iniūria cārī** the wrong of (= done to) his beloved head (= person).
25	**Hesperiae** of the Western Land = of Italy. **fraudō** (+ ablative) cheat of. **fātālibus arvīs:** i.e. the land fated to be his.
26	**interpres** the messenger.
27	**testor utrumque caput** I call both of our heads (= both of us) to witness. **mandāta** commands.
28	**manifestō in lūmine** in the clear light.
29	**hausī** I drank in, heard.
30	**tuīs . . . querēlīs** with your complaints.
31	**sponte** of my own accord.
32	**lēnīre** to soothe. **(eam) dolentem.**
33	**sōlandō** by consoling her.
34	**multa gemēns** sighing much. **animum labefactus** shaken to the heart.
35	**iussa . . . dīvum exsequitur** he carries out the commands of the gods. **revīsit** goes back to.
36	**Teucrī incumbunt** the Trojans get down to work. **celsās** tall.
37	**natat uncta carīna** the well-tarred vessel is afloat (**carīna** is a poetic singular for plural).
38–9	**frondentīs** with leaves on. **īnfabricāta** unfinished. They have time neither to strip all the leaves from the branches which they use as oars nor to finish shaping the timber.
40	**migrantīs cernās** you could have seen (them) on the move.
41–2	**velut . . . cum** just as when. **ingentem . . . farris acervum** a huge heap of grain. **formīcae** ants. **populant** pillage = take as plunder. What imagery is used throughout this simile, i.e. what are the ants implicitly compared to?
43	**campīs** over the plain. **per herbās** through the grass.
44	**convectant** they carry. **calle** on a path. **pars . . . pars . . . = aliī . . . aliī . . . grandia** vast. **trūdunt** push.
45	**obnixae** with great effort. **umerīs** with their shoulders. **cōgunt** marshal.
46	**castīgant morās** they punish delays = discipline the stragglers. **sēmita** narrow path. **fervet** boils = is in a ferment of activity.

quis tibi tum, Dīdō, cernentī tālia sēnsus,
quōsve dabās gemitūs, cum lītora fervere lātē
prōspicerēs arce ex summā, tōtumque vidērēs
50 miscērī ante oculōs tantīs clāmōribus aequor!
improbe Amor, quid nōn mortālia pectora cōgis!
īre iterum in lacrimās, iterum temptāre precandō
cōgitur et supplex animōs summittere amōrī,
nē quid inexpertum frūstrā moritūra relinquat.

*Dido sends her sister Anna down to the harbour to plead with Aeneas
to stay just long enough to enable her to come to terms with his
departure. Aeneas, however, stands firm in his resolve to leave.*

*Dido is now utterly determined to die. She pretends that a priestess
has told her that she can find release from her love if she builds a great
funeral pyre and puts on it her marriage bed and the arms and clothes
which Aeneas has left behind. These, she claims, she must burn.*

*Mercury now appears in a dream to Aeneas and tells him to leave
at once. When she sees the Trojan ships sailing away, Dido calls down a
terrible curse upon Aeneas. She says:*

55 'Sōl, quī terrārum flammīs opera omnia lūstrās,
tūque hārum interpres cūrārum et cōnscia Iūnō,
nocturnīsque Hecatē triviīs ululāta per urbēs
et Dīrae ultrīcēs et dī morientis Elissae,
accipite haec, meritumque malīs advertite nūmen
60 et nostrās audīte precēs. sī tangere portūs
īnfandum caput et terrīs adnāre necesse est,
et sīc fāta Iovis poscunt, hic terminus haeret,
at bellō audācis populī vexātus et armīs,
fīnibus extorris, complexū āvulsus Iūlī,
65 auxilium implōret videatque indigna suōrum
fūnera; nec, cum sē sub lēgēs pācis inīquae
trādiderit, rēgnō aut optātā lūce fruātur,
sed cadat ante diem mediāque inhumātus harēnā.
haec precor, hanc vōcem extrēmam cum sanguine fundō.
70 tum vōs, ō Tyriī, stirpem et genus omne futūrum
exercēte odiīs, cinerīque haec mittite nostrō
mūnera. nūllus amor populīs nec foedera suntō.
exoriāre aliquis nostrīs ex ossibus ultor
quī face Dardaniōs ferrōque sequāre colōnōs,
75 nunc, ōlim, quōcumque dabunt sē tempore vīrēs.
lītora lītoribus contrāria, fluctibus undās
imprecor, arma armīs: pugnent ipsīque nepōtēsque.'

Hannibal

154

47	*quis tum sēnsus (fuit) tibi, Dīdō, cernentī tālia.*
48	**gemitūs** sighs.
49–50	**prōspicerēs** you looked out at.... **tōtum ... miscērī ... aequor** the whole sea mixed up = the whole sea a mass of confusion.
51	**improbe** cruel. **quid** to what (extremes).
53	**supplex animōs summittere amōrī** to submit her (proud) spirit to her love as a suppliant (i.e. to Aeneas).
54	**inexpertum** untried. **frūstrā moritūra**: i.e. and so die in vain. If Aeneas would come back to her, she need not die.
55	**terrārum**: translate with **opera omnia**. **lūstrās** light up.
56	**interpres ... et cōnscia** mediator and witness
57	**nocturnīs Hecatē triviīs ululāta** Hecate whose name is shrieked at the crossroads at night. Hecate is a goddess of the underworld, associated with black magic. **per urbēs** through the cities = in every city.
58	**Dīrae ultrīcēs** avenging Furies. **Elissae**: i.e. of Dido.
59	**meritum malīs advertite nūmen** turn to my wrongs the divine power I deserve = turn your power upon my wrongs, the power they are entitled to.
60–1	*sī necesse est īnfandum caput tangere portūs.* **īnfandum caput** (for that) unspeakable being. Dido will not name Aeneas. **terrīs adnāre** to sail to land.
62	**hic terminus haeret** this bound stands firmly fixed, i.e. there is no possibility of passing the boundary stone set by fate.
63–8	All of these curses were fulfilled. Aeneas was harried by the Italian tribe, the Rutuli. He never saw Troy again. He left Iulus in order to seek help from Evander. He saw many of his men die. The peace he made with the Italians was far from favourable to the Trojans. He ruled his people for only three years, and he either drowned in a river or was killed in a battle and his body not recovered.
63	**at** even so. **bellō ... et armīs** by armed conflict with (hendiadys, i.e. the two nouns are used to express a single idea).
64	**fīnibus extorris** exiled from his home, i.e. Troy. **āvulsus** torn from.
65	**implōret** may he beg
65–6	**indigna suōrum/fūnera** the undeserved deaths of his own men.
66	**sub lēgēs pācis inīquae** under the conditions of an unjust peace.
67	**lūce**: i.e. the light of life.
68	**ante diem** before his day. **inhumātus** unburied. **harēnā** on the sand. She visualizes him lying shipwrecked on a beach, with no hope of being buried and thus of finding rest in the underworld.
69	**fundō** I pour forth = I utter.
70	**genus**: i.e. of the Trojans.
71	**exercēte** harass.
72	**suntō** let there be.
73	**exoriāre** (= **exoriāris**) **aliquis nostrīs ex ossibus ultor** may you arise, some avenger, from my bones.
74	**face ... ferrōque** with firebrand and the sword. **quī ... Dardaniōs ... sequāre** (= **sequāris**) **colōnōs** who may harry the Trojan settlers.
75	**ōlim** at some time. *quōcumque tempore vīrēs sē dabunt* at whatever time strength offers (itself).
76–7	**contrāria ... imprecor** I pray for ... in conflict with .... **ipsī nepōtēsque** themselves and their sons' sons.

Dido now sends for her sister.

at trepida et coeptīs immānibus effera Dīdō
sanguineam volvēns aciem, maculīsque trementīs
80 interfūsa genās et pallida morte futūrā,
interiōra domūs inrumpit līmina et altōs
cōnscendit furibunda gradūs ēnsemque reclūdit
Dardanium, nōn hōs quaesītum mūnus in ūsūs.
hīc, postquam Iliacās vestīs nōtumque cubīle
85 cōnspexit, paulum lacrimīs et mente morāta
incubuitque torō dīxitque novissima verba:
'dulcēs exuviae, dum fāta deusque sinēbat,
accipite hanc animam mēque hīs exsolvite cūrīs.
vīxī et quem dederat cursum fortūna perēgī,
90 et nunc magna meī sub terrās ībit imāgō.
urbem praeclāram statuī, mea moenia vīdī,
ulta virum poenās inimīcō ā frātre recēpī,
fēlīx, heu nimium fēlīx, sī lītora tantum
numquam Dardaniae tetigissent nostra carīnae.'
95 dīxit, et ōs impressa torō 'moriēmur inultae,
sed moriāmur' ait. 'sīc, sīc iuvat īre sub umbrās.
hauriat hunc oculīs ignem crūdēlis ab altō
Dardanus, et nostrae sēcum ferat ōmina mortis.'

Dido on her pyre

dīxerat, atque illam media inter tālia ferrō
100 conlāpsam aspiciunt comitēs, ēnsemque cruōre
spūmantem sparsāsque manūs. it clāmor ad alta
ātria: concussam bacchātur Fāma per urbem.
lāmentīs gemitūque et fēmineō ululātū
tēcta fremunt, resonat magnīs plangōribus aethēr,
105 nōn aliter quam sī immissīs ruat hostibus omnis
Karthāgō aut antīqua Tyros, flammaeque furentēs
culmina perque hominum volvantur perque deōrum.

78	**trepida** trembling. **coeptīs immānibus effera** wild in her dreadful design.
79–80	**sanguineam . . . aciem** her bloodshot eyes. **maculīs trementīs/ interfūsa genās** her quivering cheeks blotched with stains (of red). **pallida** pale at.
81	**interiōra . . . līmina** inner door (poetic plural). The pyre is in a courtyard of the palace.
82	**furibunda** in a frenzy. **gradūs** the steps (to the top of the pyre). **ēnsem reclūdit** unsheathes a sword.
83	i.e. Dido had asked Aeneas for the Trojan sword but had not intended to use it for this purpose.
84	**nōtum cubīle** the familiar bed.
85	**lacrimīs et mente** in tearful thought (hendiadys, the use of two nouns to convey a single idea).
86	**incubuit torō** she lay down on the bed. **novissima** last.
87	**exuviae** things he sloughed off = things which once were his. **dulcēs . . . dum** dear, as long as.
89	*perēgī* (= I have completed) *cursum quem fortūna dederat.*
90	**sub** (+ accusative) includes the idea of motion towards. **meī . . . imāgō** the ghost of me = my ghost.
91	**statuī** I set up, founded.
92	**poenās recēpī** I exacted punishment.
93	**heu nimium fēlīx, sī . . . tantum** alas, too happy, if only . . .
94	**Dardaniae . . . carīnae** the Trojan keels (= ships). King Dardanus was the founder of the Trojan race.
95	**ōs impressa torō** pressing her face in the bed. **inultae** unavenged. Dido uses the royal 'we'.
96	**iuvat** it pleases me = I choose.
97	**hauriat** let him drink in (the sight of). **ab altō** from the high sea.
98	**Dardanus** the Trojan (= Aeneas). **ōmina**, *n.pl:* i.e. the ill omen.
100–1	**conlāpsam** (her) having fallen on. **comitēs**: her friends watch from the ground, powerless to help. **ēnsem cruōre/spūmantem** the sword foaming with blood. **sparsās manūs** (her) hands spattered (with blood).
102	**concussam . . . urbem** the stricken city. **bacchātur Fāma** Rumour runs wildly.
103	**lāmentīs gemitūque et fēmineō ululātū** with lamentations and groaning and the wailing of women.
104	**fremunt** resound. **plangōribus** with sounds of mourning. **aethēr** the high heaven.
105	**nōn aliter quam sī** not otherwise than if = just as if. **immissīs ruat hostibus** were falling, after the enemy had been sent in = had broken in.
106	**Tyros** (nominative feminine) Tyre.
107	*perque culmina hominum perque (culmina) deōrum.* **culmina** roofs.

Anna's lament.

audiit exanimis trepidōque exterrita cursū
unguibus ōra soror foedāns et pectora pugnīs
110 per mediōs ruit, ac morientem nōmine clāmat:
'hoc illud, germāna, fuit? mē fraude petēbās?
hoc rogus iste mihi, hoc ignēs āraeque parābant?
quid prīmum dēserta querar? comitemne sorōrem
sprēvistī moriēns? eadem mē ad fāta vocāssēs:
115 īdem ambās ferrō dolor atque eadem hōra tulisset.
hīs etiam struxī manibus patriōsque vocāvī
vōce deōs, sīc tē ut positā, crūdēlis, abessem?
exstīnxtī tē mēque, soror, populumque patrēsque
Sīdoniōs urbemque tuam. date vulnera lymphīs
120 abluam et, extrēmus sī quis super hālitus errat,
ōre legam.' sīc fāta gradūs ēvāserat altōs,
sēmianimemque sinū germānam amplexa fovēbat
cum gemitū atque ātrōs siccābat veste cruōrēs.

V

admoneō (2)	I advise, warn
anima, -ae, *f.*	soul, life
artus, artūs, *m.*	limb
auris, auris, *f.*	ear
dictum, -ī, *n.*	word
fās, *n.* (indeclinable)	right
nefās, *n.* (indeclinable)	wrong
foedus, foederis, *n.*	treaty, agreement
Iuppiter, Iovis, *m.*	Jupiter
necesse est	it is necessary
praeclārus-a-um	splendid, famous
rōbur, rōboris, *n.*	hard wood, timber; strength; oak
secō, secāre, secuī, sectum	I cut
tēctum, -ī, *n.*	roof, house
ulcīscor, ulcīscī, ultus sum	I avenge
umbra, -ae, *f.*	shadow, ghost
ūnā	together
vērō	in truth, indeed

108	**exanimis** out of her mind. **trepidō ... cursū** in her panic-stricken running. Anna makes straight for the scene.
109	**ōra ... foedāns** marring her face with her finger-nails: **ōra** is a poetic plural. **pugnīs** with her fists (from **pugnus**).
110	**morientem (Dīdōnem).**
111	**hoc illud ... fuit?** was this (what all) that (pretence of yours meant)? **fraude** in deceit. **petēbas:** i.e. my help.
111–12	**hoc ... hoc** the objects of **parābant** (= had in store). **rogus** funeral pyre.
113	**querar:** deliberative subjunctive. **comitem:** i.e. as your companion.
114	**eadem ... ad fāta** to the same fate (as yourself). **vocāssēs** (= **vocāvissēs**) you should have called.
115	**ferrō ... tulisset** should have taken (us both) off with the sword.
116	**struxī** did I build (the pyre). **patriōs** ancestral.
117	*ut abessem tē sīc positā, crūdēlis.* **tē sīc positā** from you, placed thus (in death).
118	**exstīnxtī** (= **exstīnxistī**) you have destroyed.
119–20	*date (ut) abluam* grant that I may wash = let me wash. **lymphīs** with water. *sī quis extrēmus hālitus* if any last breath. **super** left over, i.e. still.
121	**legam** let me catch it. **fāta** having spoken. **ēvāserat** she had climbed.
122	**sēmianimem sinū germānam amplexa fovēbat** embracing her still-living (literally, half alive) sister in her arms, she caressed her.
123	**gemitū** with a groan. **ātrōs siccābat ... cruōrēs** she tried to staunch the dark blood.

Word building

Make sure that you know the following compounds of **sistō**, **sistere**, **stitī**, **statum** I stand

ab-sistō, -sistere, -stitī	I withdraw, cease from
circum-sistō, -sistere, -stetī	I surround
cōn-sistō, -sistere, -stitī	I take up position, halt
dēsistō	I leave off, cease
exsistō	I emerge, arise
obsistō + dat.	I stand in the way of, resist
resistō + dat.	I resist

Exercise 14.1

Read the following passage carefully and then answer the questions
beneath it

illa, gravīs oculōs cōnāta attollere, rūrsus
dēficit; īnfīxum strīdit sub pectore vulnus.
ter sēsē attollēns cubitōque adnīxa levāvit,
ter revolūta torō est oculīsque errantibus altō
5 quaesīvit caelō lūcem ingemuitque repertā.
    tum Iūnō omnipotēns longum miserāta dolōrem
difficilīsque obitūs Irim dēmīsit Olympō
quae luctantem animam nexōsque resolveret artūs.
nam quia nec fātō meritā nec morte perībat,
10 sed misera ante diem subitōque accēnsa furōre,
nōndum illī flāvum Prōserpina vertice crīnem
abstulerat Stygiōque caput damnāverat Orcō.
ergō Iris croceīs per caelum rōscida pennīs
mīlle trahēns variōs adversō sōle colōrēs
15 dēvolat et suprā caput astitit. 'hunc ego Dītī
sacrum iussa ferō tēque istō corpore solvō.'
sīc ait et dextrā crīnem secat: omnis et ūnā
dīlāpsus calor atque in ventōs vīta recessit.

Iris

1 **illa**: i.e. Dido.
2 **dēficit** she failed.   **īnfīxum** driven in, deep.   **strīdit** hissed.
3 **cubitō adnīxa levāvit** supporting herself on her elbow, she lifted herself up.
4 **revolūta torō est** she fell back on the bed.
5 **ingemuit** groaned deeply.   **(lūce) repertā.**
6 **omnipotēns** all-powerful.   **miserāta** pitying.
7 **difficilīs obitūs** her difficult death. Iris, Juno's messenger, is the goddess of the rainbow.
8 **luctantem animam** (her) struggling spirit.   **nexōs . . . artūs** close-locked limbs.
9 *nec meritā morte*.
10 **ante diem** before her time.   **accēnsa** on fire with.
11–12 **illī flāvum . . . vertice crīnem/abstulerat** had taken away from her the golden locks on her head = had taken a golden lock from her hair. The goddess Proserpina would cut off a lock of hair from someone who died normally, as if from an animal to be sacrificed.   **Stygiō caput damnāverat Orcō** had condemned her life to Stygian Orcus.   **Orcus:** another name for Dis or Pluto, the king of the Underworld.
13 **ergō** for this reason.   **croceīs . . . rōscida pennīs** dewy on her saffron wings = with dew on her saffron wings.
14 **adversō sōle** as the sun shone on her.
15 **suprā** above.   **hunc** i.e. the lock of hair.
16 **iussa** (as I have been) ordered.
17–18 **ait** she spoke.   *et ūnā omnis calor dīlāpsus (est)* and together (with the lock) all heat slipped from her.

1 Describe the last moments of Dido.
2 Translate ll.6–8.
3 Why is Iris sent down?
4 Why has Proserpina not taken a lock of Dido's hair?
5 Describe Iris as Virgil shows her here.
6 What does Iris do to Dido? What is the result of this action?
7 'After the unbearable climax, the book ends in tranquillity.'
 Do you feel that this is the 'right' ending? Give reasons for
 your answer.
8 Scan ll.1–2.
9 Read ll.1–5 several times. What rhythmical effect does Virgil
 achieve by having no sense pause at the end of the first line? How
 does the rhythm of these lines serve to express the sense?

Exercise 14.2

Translate into Latin
1 If you leave this land, I shall die.
2 Although Aeneas loved the queen, he
 very much wanted to sail to Italy.
3 Anna did not know why Dido had
 built the pyre.

4 Dido was so unhappy that she killed
 herself.
5 Who was not heard about the love of
 Dido and Aeneas?

Exercise 14.3

Translate into Latin
When Aeneas went down to the Underworld, he travelled through the
wood in which the souls of those who have killed themselves wander.
Among them in the shadows he saw queen Dido – or thought he saw
her. 'Unhappy Dido,' he said, 'was I the cause of your death? I swear
by (**per** + acc.) the gods that I left your country against my will. Speak
to me.' But Dido held her eyes fixed (**fīxus-a-um**) on the ground. She
was unwilling to forgive him and suddenly went away to join (she joins
= **sē coniungit cum** + abl.) her husband Sychaeus.

CHAPTER XV

The greatest war in history

In the middle of the first century BC, Cicero complained that there was no good history of Rome. Livy, the English name for Titus Livius (59 BC – AD 17), supplied this need with the monumental work to which he devoted most of his life. The 142 books of his History of Rome covered the time from the foundation of the city to 9 BC. Most of this great history has been lost.

Livy was on good terms with the Emperor, but his work is certainly not Augustan propaganda. He wrote out of a passionate love for traditional Roman values. In his account of the Second Punic War, he writes with a grim admiration of Hannibal, the great Carthaginian general who almost extinguished these values together with the city itself.

In Chapter 14, you read the terrible curses which Dido called down upon Aeneas and his descendants. She ended her speech with these dreadful words:

exoriāre aliquis nostrīs ex ossibus ultor
quī face Dardaniōs ferrōque sequāre colōnōs,
nunc, ōlim, quōcumque dabunt sē tempore vīrēs.
lītora lītoribus contrāria, fluctibus undās
imprecor, arma armīs: pugnent ipsīque nepōtēsque.

Hannibal

In our chapters from Livy's History, we see how this part of her curse was fulfilled.

Livy introduces the greatest war in history.
in parte operis meī licet mihi praefārī bellum maximē omnium
memorābile quae unquam gesta sint mē scrīptūrum, quod
Hannibale duce Carthāginiēnsēs cum populō Rōmānō gessēre.
nam neque validiōrēs opibus ūllae inter sē cīvitātēs gentēsque
5   contulērunt arma neque hīs ipsīs tantum unquam vīrium aut
rōboris fuit; et haud ignōtās bellī artēs inter sēsē sed expertās
prīmō Pūnicō cōnferēbant bellō, et adeō varia fortūna bellī fuit
ut propius perīculum fuerint quī vīcērunt. odiīs etiam prope
maiōribus certārunt quam vīribus, Rōmānīs indignantibus quod
10 victōribus victī ultrō īnferrent arma, Poenīs quod superbē
avārēque crēderent imperitātum victīs esse.

1	**in parte**: i.e. in this part. **praefārī** to state first.
1–2	**bellum maximē . . . memorābile** the most remarkable war.
2	**mē scrīptūrum (esse)**: indirect statement after **praefārī (scrībō** write about, tell the story of).
3	**gessēre = gessērunt**.
4	**validiōres opibus** stronger in resources.
5–6	**neque hīs ipsīs tantum unquam vīrium aut rōboris fuit** and not (even) was there ever so much strength and vigour to themselves = and the Romans and Carthaginians themselves were at the peak of their power.
6–7	**haud ignōtās bellī artēs . . . cōnferēbant** they brought together not unknown skills of warfare = they were by no means strangers to each other's methods of fighting. **expertās** practised. **prīmō Pūnicō bellō**: the first Punic War lasted from 264 to 241 BC. The Romans had won the war and gained, in Sicily, the beginnings of an overseas empire.
8	**propius perīculum** rather close to defeat. **quī vīcērunt = eī quī vīcērunt**, i.e. the Romans. **prope** almost.
9	**certārunt = certāvērunt**. **indignantibus**: **indignor**(1) I am indignant that.
10	**ultrō** unprovoked. **Poenī** Carthaginians.
10–11	**superbē avārēque . . . imperitātum victīs esse** that when they had been conquered they had been treated (literally, ordered about) with lordly arrogance and greed. **imperitātum esse**: impersonal use of passive.

Carthage

The young Hannibal swears an oath of undying hatred against the Romans.

    fāma est etiam Hannibalem annōrum fermē novem,
pueclīliter blandientem patrī Hamilcarī ut dūcerētur in
Hispāniam, cum exercitum eō trāiectūrus sacrificāret, altāribus
15   admōtum, tāctīs sacrīs, iūre iūrandō adāctum sē cum prīmum
posset hostem fore populō Rōmānō.

    mors Hamilcaris peropportūna et pueritia Hannibalis
distulērunt bellum. medius Hasdrubal inter patrem ac fīlium
octō fermē annōs imperium obtinuit.

Livy describes how Hannibal was greeted in Spain when he joined Hasdrubal there.

20   missus Hannibal in Hispāniam prīmō statim adventū
omnem exercitum in sē convertit; Hamilcarem iuvenem
redditum sibi veterēs mīlitēs crēdere; eundem vigōrem in vultū
vimque in oculīs intuērī. dein brevī effēcit ut pater in sē
minimum mōmentum ad favōrem conciliandum esset.

The character of Hannibal.

25   nunquam ingenium idem ad rēs dīversissimās, pārendum
atque imperandum, habilius fuit. itaque haud facile discernerēs
utrum imperātōrī an exercituī cārior esset; neque Hasdrubal
alium quemquam praeficere mālle ubi quid fortiter ac strēnuē
agendum esset, neque mīlitēs aliō duce plūs cōnfīdere aut
30   audēre. plūrimum audāciae ad perīcula capessenda, plūrimum
cōnsiliī inter ipsa perīcula erat. nūllō labōre aut corpus fatīgārī
aut animus vincī poterat. calōris ac frīgoris patientia pār; cibī
pōtiōnisque dēsīderiō nātūrālī, nōn voluptāte modus fīnītus;
vigiliārum somnīque nec diē nec nocte discrīmināta tempora.
35   multī saepe mīlitārī sagulō opertum humī iacentem inter
custōdiās statiōnēsque mīlitum cōnspexērunt. equitum
peditumque īdem longē prīmus erat; prīnceps in proelium ībat,
ultimus cōnsertō proeliō excēdēbat.

    hās tantās virī virtūtēs ingentia vitia aequābant, inhūmāna
40   crūdēlitās, perfidia plūs quam Pūnica, nihil vērī, nihil sānctī,
nūllus deum metus, nūlla religiō. cum hāc indolē virtūtum atque
vitiōrum trienniō sub Hasdrubale imperātōre meruit, nūllā rē
quae agenda videndaque magnō futūrō ducī esset praetermissā.

*In 221 Hasdrubal was murdered and Hannibal was chosen as his
successor. According to Livy, it was already Hannibal's ambition
to conquer Italy. In 219 BC he laid siege to Saguntum, a Spanish
city which was allied to Rome, and captured it. War between
Rome and Carthage was now inevitable.*

12–13	**Hannibalem ... puerīliter blandientem** (+ dat.) Hannibal, childishly coaxing. **Hamilcar** was the leading Carthaginian general.
14	**Hispānia** Spain. **eō**: i.e. to Spain. **trāiectūrus** on the point of crossing (from Carthage to Spain) ... **altāribus** the altar (plural for singular).
15	**admōtum** agrees with **Hannibalem** (l.12). **sacrīs** the sacred objects. **iūre iūrandō adāctum (esse)** was made to swear an oath. **sē** that he (Hannibal). **cum prīmum** as soon as.
17	**peropportūna** very timely.
18	**distulērunt** postponed.
19	**obtinuit** held.
21	**in sē convertit** turned upon himself = attracted the attention of.
22	**redditum (esse)**. **crēdere** they (the old soldiers) believed (historic infinitive). **vigōrem** liveliness.
23	**intuērī** they saw (historic infinitive). **dein = deinde**. **brevī**: understand **tempore**. **in sē** in his case.
24	**minimum mōmentum** of very little importance.
26	**habilius** more adaptable. **haud facile discernerēs** one (literally, you) could not easily have distinguished.
28	**quemquam** anyone. **praeficere** to put in command. **mālle** preferred (historic infinitive).
28–9	**ubi quid ... esset** whenever anything was ...
29–30	**cōnfīdere, audēre**: historic infinitives. **capessenda: capessō** I grasp, grapple with.
31	**erat**: supply **eī** there was to him = he had.
32	**calōris** of heat. **patientia** endurance: supply **erat**.
32–33	**cibī pōtiōnisque ... modus fīnītus** his appetite for food and drink was limited .... **dēsīderiō nātūrālī** by natural desire.
34	**vigiliārum** and **somnī** are dependent on **tempora** (**vigiliae**, *f. pl.* waking). **discrīmināta (sunt)** were marked off.
35	**mīlitārī sagulō** with a military cloak. **opertum** (him) covered.
35–6	**inter custōdiās statiōnēsque** amid the guard-posts and the sentry-stations (i.e. not in the **praetōrium**).
37	**īdem** the same man, i.e. in the same way, he (Hannibal) ....
38	**cōnsertō proeliō** when battle was joined.
39	**aequābant** equalled. **inhūmāna** inhuman.
40	**perfidia** treachery. **Pūnica** Carthaginian. (**Pūnica fidēs** was a proverbial expression for 'treachery'). **nihil vērī** nothing of truth. **nihil sānctī** nothing of sanctity: i.e. he showed no sense of truth and no sense of sanctity.
41	**cum hāc indole** with this character.
42	**trienniō** for three years. **meruit** he served.
43	**praetermissā: praetermittō** I omit.

Saguntum

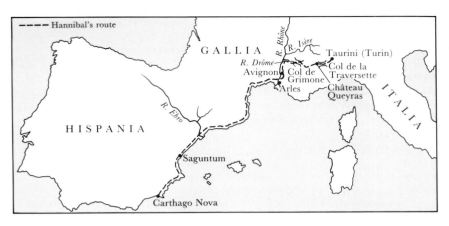

Hannibal gives his Spanish troops leave to go home for the winter.

Hannibal Saguntō captō Carthāginem Novam in hīberna
45 concesserat, ibique audītīs quae Rōmae quaeque Carthāgine
ācta dēcrētaque forent, sēque nōn ducem sōlum sed etiam
causam esse bellī, partītīs dīvenditīsque reliquiīs praedae, nihil
ultrā differendum ratus, Hispānī generis mīlitēs convocat.
 'crēdō ego vōs,' inquit, 'sociī, et ipsōs cernere pācātīs
50 omnibus Hispāniae populīs aut fīniendam nōbīs mīlitiam
exercitūsque dīmittendōs esse aut in aliās terrās trānsferendum
bellum. itaque cum longinqua ab domō īnstet mīlitia
incertumque sit quandō domōs vestrās vīsūrī sītis, si quis vestrum
suōs invīsere vult, commeātum dō. prīmō vēre ēdīcō adsītis, ut
55 dīs bene iuvantibus bellum ingentis glōriae praedaeque futūrum
incipiāmus.' omnibus ferē vīsendī domōs oblāta ultrō potestās
grāta erat. per tōtum tempus hiemis quiēs inter labōrēs aut iam
exhaustōs aut mox exhauriendōs renovāvit corpora animōsque
ad omnia dē integrō patienda; vēre prīmō ad ēdictum convēnēre.

*In May 218 BC, Hannibal crossed the Ebro. In July and August he
crossed the Pyrenees and was then confronted with the wide river
Rhône.*

*Hannibal sends a detachment of his men to cross over the river
upstream and attack the enemy on the far bank from behind.*

60 iamque omnibus satis comparātīs ad trāiciendum terrēbant ex
adversō hostēs omnem rīpam equitēs virīque obtinentēs. quōs ut
āverteret, Hannōnem vigiliā prīmā noctis cum parte cōpiārum,
maximē Hispānīs, adversō flūmine īre iter ūnius diēī iubet et, ubi
prīmum possit quam occultissimē trāiectō amnī, circumdūcere
65 agmen ut adoriātur ab tergō hostēs. ad id datī ducēs Gallī
ēdocent inde mīlia quīnque et vīgintī fermē suprā parvae īnsulae
circumfūsum amnem minus altō alveō trānsitum ostendere.

45	**audītīs quae . . .** = **eīs** (those things) **audītīs quae . . . .**
46	**ācta dēcrētaque forent:** translate **forent** as **essent.** **sēque . . . ducem . . . esse** and (hearing) that he was the leader . . .
47	**partītīs dīvenditīsque . . . reliquiīs** having divided out the rest or sold it piecemeal.
47–8	**nihil ultrā differendum (esse) ratus** thinking that he should postpone matters no further.
49–50	**et ipsōs** yourselves (also), i.e. as I do. **pācātīs . . . populīs** the tribes having been won over.
50	**omnibus Hispāniae populīs:** Hannibal had in fact conquered only the hostile Spanish tribes who lived South of the Ebro. **mīlitiam** campaign.
52	**īnstet** is at hand (subjunctive after **cum** since).
54	**commeātum** military leave, furlough. **ēdīcō adsītis:** supply **ut** I order you to be here.
55	**bellum ingentis glōriae praedaeque futūrum** a war which will be one of (i.e. bringing) great glory and great booty.
56	**ferē** almost. **ultrō** unasked. **vīsendī . . . potestās** opportunity of seeing.
57–8	**aut iam exhaustōs aut mox exhauriendōs** which they had already endured and which they soon had to endure.
59	**dē integrō** afresh. **ad ēdictum** in accordance with his command. **convēnēre** = **convēnērunt.**
60–1	**ex adversō** on the other side.
61	**virī** = **peditēs.**
62	**āverteret** he might draw them off. **Hannōnem:** the nominative of this name is Hanno.
63	**adversō flūmine** with the river against them = upstream.
64	**quam occultissimē** as secretly as possible. **trāiectō amnī** after crossing the river.
65	**adoriātur** he could attack. **ad id datī** appointed for this purpose.
66–7	**ēdocent . . . amnem . . . trānsitum ostendere** told them that . . . the river . . . afforded a crossing. **suprā** above = upstream. **minus altō alveō** because of its shallower channel.

parvae īnsulae circumfūsus amnis   The river Rhône at Avignon

ibi raptim caesa māteria ratēsque fabricātae in quibus equī
virīque et alia onera trāicerentur. Hispānī sine ūllā mōle flūmen
70 trānāvēre. et alius exercitus ratibus iūnctīs trāiectus, castrīs
prope flūmen positīs, nocturnō itinere atque operis labōre fessus
quiēte ūnius diēī reficitur, intentō duce ad cōnsilium opportūnē
exsequendum. posterō diē profectī, ex locō ēditō fūmō
significant trānsīsse et haud procul abesse; quod ubi accēpit
75 Hannibal, dat signum ad trāiciendum.

(V)

concēdō, concēdere, concessī	I go away, retire, give way to	**praeter** + accusative	beside, past, along
dēcernō, dēcernere, dēcrēvī, dēcrētum	I decide, decree	**strēnuus-a-um**	active, vigorous
efficiō, efficere, effēcī effectum ut + subjunctive	I bring it about that	**trāiciō, trāicere, trāiēcī, trāiectus**	I cross over; I send (a thing or a person) across
fatīgō (1)	I tire out, harass	**usquam**	anywhere
fermē	almost, about	**nusquam**	nowhere
hīberna, -ōrum, *n.pl.*	winter quarters	**vitium, vitiī,** *n.*	vice, fault
longinquus-a-um	distant		
memorābilis-e	remarkable, memorable		

(G) Word building

Make sure that you know the following compounds of **currō**, **currere**, **cucurrī**, **cursum**

ac-currō, -currere, -currī, -cursum	I run to	**incurrō**	I run to; I attack
		percurrō	I run through
concurrō	I run together; I meet	**praecurrō**	I run before
dēcurrō	I run down	**prōcurrō**	I run forward
discurrō	I run about	**recurrō**	I run back
excurrō	I run out	**succurrō** + dative	I run to help

(G) The historic infinitive

The present infinitive is frequently used in place of a past tense of the
indicative to describe exciting or striking actions or emotions. We call
this the 'historic infinitive'.

In the following excerpt from Livy's character sketch of
Hannibal, the infinitives convey how remarkable it was that one so
young should be given such responsibility and should inspire so much
confidence and boldness:

neque Hasdrubal alium quemquam praeficere mālle ubi quid
fortiter ac strēnuē agendum esset, neque mīlites aliō duce plūs
cōnfīdere aut audēre.

68	**raptim** hastily. **caesa (est).** **māteria** timber. **ratēs fabricātae** rafts were constructed.

68	**raptim** hastily. **caesa (est).** **māteria** timber. **ratēs fabricātae** rafts were constructed.
69	**sine ūlla mōle** without any trouble, difficulty.
70	**trānāvēre** swam across. **alius exercitus** the rest of the army. **ratibus iūnctīs** on the rafts which they had made.
71	**nocturnō itinere** by the night march. **operis** of the construction-work, i.e. the raft-building.
72	**reficitur** was restored, rested.
72	**intentō duce ad** the general (being) intent on . . . , eager to . . . . **opportūnē** at the right time.
73	**exsequendum: exsequor** I carry out. **ex locō ēditō** from a high place. **fūmō** by a smoke signal.
74	**trānsīsse . . . abesse** understand **sē** as the subject of these infinitives.

G 'ēre' for 'ērunt' (3rd person plural, perfect indicative active)

convēn-ēre
trānāv-ēre

These are alternative forms for **convēn-ērunt** and **trānāv-ērunt**. Such forms are commonly used, especially in poetry.

G Comparative clauses

Hannibal tam fortis est quam pater (erat)
Hannibal is as brave as his father was.

mīlitēs flūmen trānsiērunt sīcut iussī erant
The soldiers crossed the river as they had been ordered.

quantō erat in diēs gravior pugnātiō, tantō crēbriōres nūntiī ad Caesarem mittēbantur
The more severe the fighting grew from day to day, the more frequently messages were sent to Caesar.

Learn the following comparative conjunctions

sīcut	just as, just as if
aequē ac	as much as
alius-a-ud ac	different from
īdem, eadem, idem ac	the same as
quasi	as if
quō . . . eō	the more . . . the more
tālis . . . quālis	such . . . as
tantus . . . quantus	as much (great) . . . as
tot . . . quot	as many . . . as
tam . . . quam	as (so) . . . as
ut . . . ita	as . . . so
ut . . sīc	as . . . so

Exercise 15.1

Translate

1 quō propius ea contentiō accēdit, eō clārius id perīculum appāret.
2 flūmen trānsīte, ō mīlitēs, sīcut imperāvī.
3 pater idem cōnsilium mihi ac tibi dedit.
4 tālis est quālem tū eum esse scrīpsistī.
5 ut Cicerō mihi imperāvit, ita faciam.
6 quot hominēs, tot sententiae.
7 quantō fortiōrem Hannibal sē praebēbat, tantō magis mīlitēs eum amābant.
8 tam fēlīx ōlim fuī quam fīlius tuus nunc est.
9 tantum theātrum quantum Pompēius aedificāvit numquam anteā vīdī.

Exercise 15.2

Read the following passage carefully and then answer the questions beneath it

    Hannibal, nāvium agmen, ad excipiendum adversī impetum flūminis, parte superiōre trānsmittēns tranquillitātem īnfrā trāicientibus lintribus praebēbat. Gallī occursant in rīpā cum variīs ululātibus cantūque mōris suī, quatientēs scūta super
5   capita vibrantēsque dextrīs tēla, quamquam ex adversō terrēbat tanta vīs nāvium cum ingentī sonō flūminis et clāmōre variō nautārum mīlitumque, et quī nītēbantur perrumpere impetum flūminis et quī ex alterā rīpā trāicientēs suōs hortābantur.
    iam satis paventēs adversō tumultū terribilior ab tergō
10 adortus clāmor, castrīs ab Hannōne captīs. mox et ipse aderat ancepsque terror circumstābat, et ē nāvibus tantā vī armātōrum in terram ēvādente et ab tergō imprōvīsā premente aciē. Gallī postquam utrōque vim facere cōnātī pellēbantur, quā patēre vīsum maximē iter perrumpunt trepidīque in vīcōs passim suōs
15 diffugiunt. Hannibal, cēterīs cōpiīs per ōtium trāiectīs, spernēns iam Gallicōs tumultūs castra locat.

1–2	**nāvium agmen . . . parte superiōre trānsmittēns** sending across the column of ships upstream. **ad excipiendum adversī impetum flūminis** in order to take the force of the current (literally: the onslaught of the river against them). **tranquillitātem** calm water. **īnfrā** below = downstream.
3	**lintribus** for the small boats. Hannibal is the subject of **praebēbat**. **occursant** rushed to meet them.
4	**ululātibus** with war-whoops. **cantū** with singing. **mōris suī** of their custom = in their customary manner. **quatientēs** brandishing.
5	**vibrantēs** shaking. **dextrīs** in their right hands. **ex adversō** opposite them.
6	**sonō** by the noise.
7	**nītēbantur** were striving.
8	**trāicientēs**: agrees with **suōs**.

9 **iam satis paventēs adversō tumultū** as they (the Gauls) were already extremely frightened by the noise to their front.

9–10 *(eōs) paventēs adortus (est) clāmor:* **adortus (est)** assailed. The shouting which arose 'hit' them.

11 **anceps terror** a twofold terror.

11–12 **vī . . . ēvadente** a force coming out (of the river). **imprōvīsā . . . aciē** an unexpected line (= troops) . . . = troops unexpectedly . . .

13 **utrōque vim facere** to fight in both directions. **pellēbantur** they were being driven back.

13–14 **quā patēre vīsum (est) maximē iter** where the way seemed most to lie open = by the route which seemed to offer the best escape. **trepidī** in panic. **vīcōs** villages.

14 **passim** in different directions.

15 **per ōtium** undisturbed.

15–16 **spernēns iam Gallicōs tumultūs** now holding the sound and fury of the Gauls in contempt. **locat** pitched.

1 Why were the small vessels able to cross over in calm water?
2 Describe the actions and the state of mind of the Gauls before the shout arose behind them. How does Livy try to make his description here exciting?
3 What was the 'anceps terror'?
4 How bravely do you think that the Gauls behave? Do you share Hannibal's view of them?
5 Translate the second paragraph.
6 Basing your answer on the evidence given in this chapter, say what your opinion is of Hannibal both as a man and as a general.

Exercise 15.3

Translate into Latin

1 Hamilcar led his son to the altar before he crossed over into Italy.
2 Hannibal swore that he would always be an enemy to the Roman people.
3 When they saw Hannibal, the soldiers were very happy.
4 Nobody was a better general than Hannibal.
5 Hannibal ordered Hanno to cross the river and attack the enemy from the rear.

Exercise 15.4

Translate into Latin

When he realized that Julius Caesar wished to become king, Cassius tried to persuade his friend Brutus to kill him. Although Brutus at first delayed – for Caesar was very dear to him –, he later promised to obey Cassius's orders.

When the senate had met, Brutus and Cassius and certain other men approached Caesar and surrounded him. Then they killed him very cruelly with their daggers (**pugiō, pugiōnis**, *m.*). Caesar, while he fell dying near the statue (**statua, -ae**, *f.*) of Pompey, said, 'You too, Brutus?'

CHAPTER XVI

Hannibal reaches the Alps

Hannibal's men are confronted by the fearsome sight of the Alps.
Hannibal ab Druentiā, campestrī maximē itinere, ad Alpēs cum
bonā pāce incolentium ea loca Gallōrum pervēnit. tum ex
propinquō vīsa montium altitūdō nivēsque caelō prope
immixtae, tēcta īnformia imposita rūpibus, pecora iūmentaque
5 torrida frīgore, hominēs intōnsī et incultī, animālia inanimaque
omnia rigentia gelū terrōrem renovārunt.

*The mountain people threaten Hannibal's advance; he sends his
spies among them.*
    ērigentibus in prīmōs agmen clīvōs appāruērunt imminentēs
tumulōs īnsīdentēs montānī, quī, sī vallēs occultiōrēs īnsēdissent,
coortī ad pugnam repente, ingentem fugam strāgemque
10 dedissent.
    Hannibal cōnsistere signa iussit; Gallīsque ad vīsenda loca
praemissīs, postquam comperit trānsitum eā nōn esse, castra
quam extentissimā potest vallē locat. tum per eōsdem Gallōs,
haud sānē multum linguā mōribusque abhorrentēs, cum sē
15 immiscuissent conloquiīs montānōrum, ēdoctus interdiū tantum
obsidērī saltum, nocte in sua quemque dīlābī tēcta, lūce prīmā
subiit tumulōs, ut ex apertō atque interdiū vim per angustiās
factūrus.

1   **Druentiā**: the river Durance.   **campestrī** over the plains.   **maximē** for the most part.

2   **ea loca**, *n.pl.* those places (the object of **incolentium**).

3–6   **vīsa** agrees with **altitūdō**, the first noun in the list, but it should be taken with all the subjects of **renovārunt**: the sight of (all these things) renewed their terror.

3–4   **nivēs . . . prope immixtae** the snow (literally: the snows) almost merging with . . . .   **tēcta īnformia** shapeless huts.   **iūmenta** beasts of burden.

5   **torrida** frost-bitten.   **intōnsī et incultī** shaggy and shabby.   **inanima** inanimate objects.

6   **rigentia gelū** stiff with cold.   **renovārunt = renovāvērunt** renewed.

7   **ērigentibus in prīmōs agmen clīvōs** to them leading their column up into the first slopes = as their column began to climb the first slopes.

7–9   **montānī** mountain-dwellers (the subject of the sentence).
imminentēs tumulōs īnsīdentēs occupying the overhanging hill-tops.   **vallēs occultiōrēs** less visible valleys.   **coortī** springing out.   **strāgem** slaughter.

11   **cōnsistere signa iussit** he ordered the standards to halt = he ordered a (general) halt.   **Gallīs**: these were the Gauls serving in Hannibal's army.

12   **comperit** he discovered.   **trānsitum eā (viā) nōn esse** that there was no crossing that way = that it was impossible to cross there.

13   **quam extentissimā potest vallē** in the broadest valley he could find.

14   **haud sānē multum . . . abhorrentēs** not very different (i.e. from the Gauls of the neighbourhood).

14–5   **cum sē immiscuissent conloquiīs montānōrum** when they had mixed themselves in the conversations of the mountain people = when they had infiltrated the mountain people and picked up their conversations.

15–16   **ēdoctus interdiū tantum obsidērī saltum** having discovered that the pass was guarded only in the daytime.   **dīlābī** melted away.

17   **subiit** he marched up to.   **ut** (+ future participle) as though in order to.   **ex apertō** openly.

17–18   **vim per angustiās factūrus** to force his way through the narrow pass.

imminentēs tumulōs īnsīdentēs montānī

Hannibal's stratagem.

diē deinde simulandō aliud quam quod parābātur
20 cōnsūmptō, cum eōdem quō cōnstiterant locō castra
commūnīssent, ubi prīmum dēgressōs tumulīs montānōs
laxātāsque sēnsit custōdiās, plūribus ignibus quam prō numerō
manentium in speciem factīs impedīmentīsque cum equite
relictīs et maximā parte peditum, ipse cum expedītīs raptim
25 angustiās ēvādit iīsque ipsīs tumulīs quōs hostēs tenuerant
cōnsēdit.

prīmā deinde lūce castra mōta et agmen reliquum incēdere
coepit. iam montānī signō datō ex castellīs ad statiōnem solitam
conveniēbant, cum repente cōnspiciunt aliōs arce occupātā suā
30 super caput imminentēs, aliōs viā trānsīre hostēs. utraque simul
obiecta rēs oculīs animīsque immōbilēs parumper eōs dēfīxit;
deinde, ut trepidātiōnem in angustiīs suōque ipsum tumultū
miscērī agmen vīdēre, quidquid adiēcissent ipsī terrōris satis ad
perniciem fore ratī, dīversīs rūpibus dēcurrunt.

A narrow escape.
35 tum vērō simul ab hostibus, simul ab inīquitāte locōrum
Poenī oppugnābantur, plūsque inter ipsōs, sibi quōque tendente
ut perīculō prius ēvāderet, quam cum hostibus certāminis erat. et
equī maximē īnfestum agmen faciēbant, quī et clāmōribus
dissonīs, quōs nemora etiam repercussaeque vallēs augēbant,
40 territī trepidābant, et ictī forte aut vulnerātī adeō
cōnsternābantur, ut strāgem ingentem simul hominum ac
sarcinārum omnis generis facerent; multōsque turba, cum
praecipitēs utrimque angustiae essent, in immēnsum altitūdinis
dēiēcit; et iūmenta cum oneribus dēvolvēbantur.
45 quae quamquam foeda vīsū erant, stetit parumper tamen
Hannibal ac suōs continuit, nē tumultum ac trepidātiōnem
augēret; deinde, postquam interrumpī agmen vīdit, dēcurrit ex
superiōre locō et, cum impetū ipsō fūdisset hostem, suīs quoque
tumultum auxit. sed is tumultus mōmentō temporis, postquam
50 līberāta itinera fugā montānōrum erant, sēdātur, nec per ōtium
modo sed prope silentiō mox omnēs trāductī.

castellum inde, quod caput eius regiōnis erat, vīculōsque
circumiectōs capit et captō cibō ac pecoribus per trīduum
exercitum aluit; et, quia nec montānīs prīmō perculsīs nec locō
55 magnō opere impediēbantur, aliquantum eō trīduō viae cōnfēcit.

19–20	**diē . . . cōnsūmptō** spending a day (ablative absolute). **simulandō** in pretending (to do). **aliud quam quod parābātur** something other than what was being prepared.
22	**laxātās . . . custōdiās** they had relaxed their vigilance.
22–3	**plūribus ignibus quam prō numerō manentium** more fires than in proportion to (= were needed by) the number of those who stayed, i.e. he lit more fires than necessary.
23	**in speciem** as a pretence. **impedīmentīs: impedimenta,** *n.pl.* baggage.
24	**cum expedītīs** with the light-armed soldiers.
24–5	**raptim angustiās ēvādit** speedily emerged at the top of the pass.
27	**mōta (sunt)**. **incēdere** to advance.
28	**ex castellīs** from (their) fortified villages. **statiōnem solitam** their customary watch-post.
29–30	**aliōs . . . aliōs . . . hostēs**: i.e. they saw some of their enemy in one place, others in another.
31	**obiecta** presented to. **immōbilēs . . . eōs dēfīxit** fixed them motionless to the spot. **parumper** for a little = for a short while.
32	**trepidātiōnem** panic, i.e. of Hannibal's men.
32–3	**suōque ipsum tumultū miscērī agmen** and that the column was being thrown into disorder by its own confusion.
33–4	**quidquid adiēcissent ipsī terrōris satis ad perniciem fore ratī** thinking that whatever terror (**quidquid terrōris**) they themselves could add would be sufficient to destroy them. **dīversīs** on both sides.
35	**ab inīquitāte locōrum** by the unfavourable nature of the place = by the unfavourable terrain. The ground was uneven, with many defiles and no room to deploy.
36–7	**plūs . . . certāminis erat** there was more of a struggle. **sibi quōque tendente** as each man strove for himself. **prius** first.
38	**maximē īnfestum** most disordered.
39	**dissonīs** discordant. **nemora,** *n.pl.* the woods. **repercussae** echoing (agreeing with both **nemora** and **vallēs**).
40	**trepidābant** began to panic. **ictī** being hit.
41	**cōnsternābantur: cōnsternō**(1) I alarm. **strāgem** destruction.
42	**sarcinārum,** *f.pl.* of baggage. **turba** the confusion.
43	**utrimque** on both sides. **in immēnsum altitūdinis** to an enormous depth.
45	**foeda vīsū** dreadful to behold. **parumper** for a short while.
46	**continuit** held back. **trepidātiōnem** panic.
47	**interrumpī** was being broken apart.
48	**cum** although.
50	**līberāta** cleared. **sēdātur** was calmed. **per ōtium** without harrassment.
51	**trāductī (sunt)** were led across.
52–3	**vīculōs circumiectōs** the outlying villages. **trīduum** a period of three days.
54	**aluit: alō, alere, aluī** I feed. **montānīs prīmō perculsīs** by the mountain people, utterly demoralized at the outset. **locō** by the (nature of the) country.
55	**magnō opere** = **magnopere**. **aliquantum . . . viae** some distance. **eō trīduō** in those three days.

The treachery of the Gauls leads the Carthaginians into another
extremely dangerous situation.

perventum inde ad frequentem cultōribus alium populum.
ibi nōn bellō apertō sed suīs artibus, fraude et īnsidiīs, est prope
circumventus. magnō nātū prīncipēs castellōrum ōrātōrēs ad
Poenum veniunt, aliēnīs malīs doctōs memorantēs amīcitiam
60 mālle quam vim experīrī Poenōrum; itaque oboedienter
imperāta factūrōs; commeātum itinerisque ducēs et ad fidem
prōmissōrum obsidēs acciperet.

Hannibal nec temere crēdendum nec aspernandōs ratus, nē
repudiātī apertē hostēs fierent, benignē cum respondisset,
65 obsidibus quōs dabant acceptīs et commeātū quem in viam ipsī
dētulerant ūsus, compositō agmine ducēs eōrum sequitur.
prīmum agmen elephantī et equitēs erant; ipse post cum rōbore
peditum circumspectāns sollicitus omnia incēdēbat.

ubi in angustiōrem viam et ex parte alterā subiectam iugō
70 ventum est, undique ex īnsidiīs barbarī ā fronte ab tergō coortī,
comminus ēminus petunt, saxa ingentia in agmen dēvolvunt.
maxima ab tergō vīs hominum urgēbat. tunc quoque ad
extrēmum perīculī ac prope perniciem ventum est; nam dum
cūnctātur Hannibal dēmittere agmen in angustiās, occursantēs
75 per oblīqua montānī interruptō mediō agmine viam īnsēdēre,
noxque ūna Hannibalī sine equitibus atque impedīmentīs ācta
est.

arx, arcis, *f.*	citadel
cōnsīdō, cōnsīdere, cōnsēdī	I settle in, occupy
īnsīdō, īnsīdere, īnsēdī	I sit on, occupy
fraus, fraudis, *f.*	deceit
fundō, fundere, fūdī, fūsum	I rout, cause to flee; I pour
impedīmenta,	
impedīmentōrum *n.pl.*	baggage
īnfestus-a-um	dangerous, hostile; endangered
interdiū	by day
nēquīquam	in vain, to no purpose
nix, nivis, *f.*	snow
pecus, pecoris, *n.*	flock
propinquus-a-um	near
rūpēs, rūpis, *f.*	rock, cliff, crag
vallēs, -is, *f.*	valley
tumultus, -ūs, *m.*	confusion, chaos, revolt

56	**perventum (est)**: impersonal use of the passive, 'they came'. **ad frequentem cultōribus ... populum** to a people (= a settlement) crowded with inhabitants = to a well-populated settlement.				
57	**suīs artibus** by his (i.e. Hannibal's) own devices. This is a reference to the treacherous reputation of the Carthaginians. **prope** nearly.				
58	**circumventus** outmanœuvred. **magnō nātū prīncipēs** the elder headmen. **ōrātōrēs** as speakers = as a deputation.				
59	**Poenum** the Carthaginian, i.e. Hannibal. **aliēnīs malīs** by other men's evils. **memorantēs** saying (take before **aliēnīs malīs**)				
60	**mālle**: understand **sē** before **mālle**. **experīrī** to make trial of. **oboedienter** obediently.				
61	**factūrōs (esse)**. **commeātum** food-supplies.				
61–2	**ad fidem prōmissōrum** as a pledge that they would fulfil their promises.				
62	**acciperet** let him accept.				
63	**nec temere crēdendum nec aspernandōs ratus** thinking that they should neither be blindly trusted nor treated with contempt.				
64	**repudiātī: repudiō**(1) I reject. **benignē** kindly.				
66	**compositō agmine** with his column formed = he formed his column and. . . .				
68	**circumspectāns sollicitus omnia** looking round anxiously at everything.				
69	**ex parte alterā subiectam iugō** overhung on one side by a ridge: **subiectam** (= overhung) agrees with **viam**.				
70	**ā fronte ab tergō** from the front and the rear.				
71	**comminus ēminus** at close quarters and at long range. **petunt** they attack.				
72	**vīs hominum** force (= number) of men. **tunc quoque** on this occasion too.				
73	**prope perniciem** near to destruction.				
74–5	**occursantēs per oblīqua montānī** the mountain people, making an attack on the flank.				
75	**interruptō: interrumpō** I break through.				
76	**Hannibalī** by Hannibal (dative of the agent).				

Exercise 16.1

Give the meaning of the following words

agō	āctor	āctus, -ūs	āctiō	āctum, -ī, *n*.	agmen
canō	cantor	cantus	cantiō		carmen
certō			certātiō		certāmen
currō	cursor	cursus			
fluō		fluctus			flūmen
imperō	imperātor			imperātum	imperium
incendō					incendium
spectō	spectātor		spectātiō		spectāculum
scrībō	scrīptor		scrīptiō	scrīptum	

G 'quisque', 'quis', 'quisquam'

1 quisque, quaeque, quidque each man, woman or thing

It is used *with superlatives:*
optimus quisque each best man, i.e. all the best men

with ordinals:
decimus quisque each tenth man, i.e. every tenth man

with sē and *suus:*
nocte in sua tēcta quisque dēlāpsus est
At night each man (i.e. all of them) slipped away to his hut.

prō sē quisque pugnāvit
Each man fought for himself.

2 quis, qua, quid anyone, anything

It is used after **sī, nisi, num** and **nē**:
sī quis Hannibalem oppugnat, vincitur
If anyone attacks Hannibal, he is beaten.

rogāvī num quis Hannibalem oppugnāre audēret
I asked whether anyone would dare to attack Hannibal.

3 quisquam, quidquam/quicquam anyone, anything

It is used after a negative word or **vix** or **aegrē**
(scarcely):
vix quisquam Hannibalem oppugnāre audēbit
Scarcely anyone will dare to attack Hannibal.

Apart from their nominative singular and accusative neuter singular, these words decline like **quī, quae, quod**.

Exercise 16.2

Translate

1 haud quemquam vīdī laetiōrem quam tē.

2 Hannibal nocte cum cōpiīs prōcessit nē quis eās vidēret.

3 num quis tam stultus est ut Alpēs trānsīre cōnētur?

4 rogāvī num quem Gallī cēpissent.

5 vix quisquam putāvit Hannibalem Alpēs trānsīre posse.

6 tam malus est ut optimum quemque vehementer ōderit.

7 sī quis hoc dīxit, maximē errāvit.

8 imperātor, quī mīlitēs ignāvōs pūnīre volēbat, decimum quemque interfēcit.

Exercise 16.3

Translate into English

posterō diē, iam sēgnius incursantibus barbarīs, iūnctae cōpiae
saltusque haud sine clādē, maiōre tamen iūmentōrum quam hominum
perniciē, superātus. inde montānī pauciōrēs iam et latrōciniī magis
quam bellī mōre concursābant modo in prīmum, modo in novissimum
5   agmen, utcumque aut locus opportūnitātem daret aut prōgressī
morātīve aliquam occāsiōnem fēcissent. elephantī sīcut per artās viās
magnā morā agēbantur, ita tūtum ab hostibus quācumque incēderent,
quia īnsuētīs adeundī propius metus erat, agmen praebēbant.

1	**sēgnius incursantibus barbarīs** the barbarians attacking less vigorously. **iūnctae (sunt):** i.e. the cavalry and baggage train rejoined the rest of the force.
2	**saltus** the pass. **iūmentōrum** of baggage animals.
3	**superātus (est)** was topped: i.e. they reached the top of the pass. **latrōciniī** of brigandage (dependent, like **bellī**, on **mōre**).
4	**concursābant** were skirmishing. **modo . . . modo** sometimes . . . sometimes. **novissimum** rear.
5	**utcumque** in whatever way. **opportūnitātem** a chance.
6	**-ve** or. It is attached to the second of the two alternatives (compare **-que**), e.g. **bonus malusve** = good or bad.
6–7	**sīcut . . . ita** just as . . . so = even though . . . still. **artās** narrow.
7–8	**tūtum . . . agmen praebēbant (praebēbant** = made). **quācumque incēderent** wherever they went. **īnsuētīs** (to their enemies who were) unaccustomed (to the elephants).

Exercise 16.4

Translate into Latin

1 Hannibal praised all the best men in his army.
2 They suffered so many dangers that they almost despaired.
3 Do not trust the Gauls. They will certainly betray you.
4 After they had fought for four hours, the Carthaginians overcame the enemy.
5 Hannibal is as brave as his father Hasdrubal.

Exercise 16.5

Translate into Latin

Hannibal was so good a general that his men trusted him very greatly.
When he had warned them about the dangers of the mountains, he
encouraged them with these words. 'You must not be afraid. You
have crossed the wide river Rhône (**Rhodanus,-ī**, *m.*); you have
already defeated the natives of this land. I do not fear that you will
now show yourselves (to be) cowards. If you fight bravely, you will
overcome the enemy.' Roused by these words, the Carthaginians
went forward cheerfully, but when at last they saw how high the Alps
were, they were terrified.

CHAPTER XVII

Hannibal crosses the Alps

The Carthaginians reach the top of the Alps.

nōnō diē in iugum Alpium perventum est per invia plēraque et
errōrēs, quōs aut dūcentium fraus aut temerē initae vallēs ā
coniectantibus iter faciēbant. bīduum in iugō statīva habita
fessīsque labōre ac pugnandō quiēs data mīlitibus; iūmenta
5   aliquot, quae prōlāpsa in rūpibus erant, sequendō vestīgia
agminis in castra pervēnēre.

fessīs taediō tot malōrum nivis etiam cāsus ingentem
terrōrem adiēcit. per omnia nive opplēta cum signīs prīmā lūce
mōtīs sēgniter agmen incēderet pigritiaque et dēspērātiō in
10   omnium vultū ēminēret, praegressus signa Hannibal in
prōmuntoriō quōdam, unde longē ac lātē prōspectus erat,
cōnsistere iussīs mīlitibus Italiam ostentat subiectōsque Alpīnīs
montibus Circumpadānōs campōs, moeniaque eōs tum
trānscendere nōn Italiae modo sed etiam urbis Rōmānae; cētera
15   plāna, prōclīvia fore; ūnō aut summum alterō proeliō arcem et
caput Italiae in manū ac potestāte habitūrōs.

The hazardous descent.

prōcēdere inde agmen coepit iam nihil nē hostibus quidem
praeter parva fūrta per occāsiōnem temptantibus. cēterum iter
multō quam in adscēnsū fuerat difficilius fuit; omnis enim fermē
20   via praeceps, angusta, lūbrica erat, ut neque sustinēre sē ab
lāpsū possent aliīque super aliōs et iūmenta in hominēs
occiderent.

ventum deinde ad multō angustiōrem rūpem atque ita rēctīs
saxīs ut aegrē expedītus mīles temptābundus manibusque
25   retinēns virgulta ac stirpēs circā ēminentēs dēmittere sēsē posset.
nātūrā locus iam ante praeceps recentī lāpsū terrae in pedum
mille admodum altitūdinem abruptus erat. ibi cum velut ad
fīnem viae equitēs cōnstitissent, mīrantī Hannibalī quae rēs
morārētur agmen nūntiātur rūpem inviam esse. dēgressus deinde
30   ipse ad locum vīsendum. haud dubia rēs vīsa quīn per invia circā
nec trīta anteā, quamvīs longō ambitū, circumdūceret agmen.

1   **nōnō diē**: i.e. from the foot of the Alps.

1–2   **per invia plēraque et errōrēs** travelling mostly through pathless tracts and by wrong routes.

2–3   **temere initae vallēs ā coniectantibus iter** valleys blindly entered by (the Carthaginians) guessing the way. **bīduum** for two days. **statīva** a camp.

7   **taediō** (+ genitive) by discouragement at. **nivis cāsus**: it was now late September.

8   **opplēta** covered.

9   **sēgniter** sluggishly. **pigritia** weariness.

10   **ēminēret** were obvious (the verb is singular in Latin because the weariness and the despair form one idea). **praegressus signa** going ahead of the standards.

11   **in prōmuntoriō** on a ledge. **prōspectus** view.

12   **ostentat** he shows.

13   **Circumpadānōs** around the river Po.

13–14 **eōs trānscendere**: accusative + infinitive: supply 'he said that' to introduce this: **moenia** is the object of **trānscendere.**

15   **plāna** flat. **prōclīvia** downhill and easy. **summum alterō** at most by a second, i.e. at most two . . .

16   **caput** capital.

17–18 **nihil nē hostibus quidem . . . temptantibus** with the enemy no longer making any assaults . . . **fūrta** stealthy attacks. **per occāsiōnem** as opportunity offered. **cēterum** but.

20   **lūbrica** slippery.

20–1 **neque . . . que:** they could not stop themselves from slipping (**ab lapsū**) but . . .

22   **occiderent** fell.

23   **ventum (est). rūpem** precipitous cliff. **rēctīs** vertical.

24   **aegrē** with difficulty. **expedītus** unencumbered. **temptābundus** feeling his way.

25   **virgulta ac stirpēs circā ēminentēs** the bushes and roots which projected all around.

26–7 **in pedum mille admodum altitūdinem** to a depth of nearly 1,000 feet. **velut** as though.

29   **inviam** impassable.

30–1 **haud dubia rēs vīsa quīn . . . circumdūceret** there seemed no doubt that he must lead round . . . . **per invia circā nec trīta anteā** through the trackless and previously untrodden areas round about. **quamvīs longō ambitū** however long the detour might prove.

But this route proved totally impassable. Above the old, untouched snow there was a new layer of moderate depth and at first they found it easy to gain a foothold as they advanced, since it was soft and not too deep. However, as it melted under the feet of so many men and animals, they had to make their way over the bare ice underneath and the liquid slush of the melting snow.

There was a grim struggle here, since the slippery path did not afford a foothold and the downward slope made their feet slide even more quickly, so that, whether they used their knees or their hands to try to get up, even these supports would slip from under them and they would fall down again. And there were no stems or roots thereabouts for them to get a hold on with their hands and feet to help them struggle up. There was only the smooth ice and the slushy snow, on which they constantly rolled about. The baggage animals sometimes even cut into the bottommost crust of snow as they went along; they slipped forward and, as they struck out strongly with their hooves, they broke right through so that a large number of them stuck fast in the hard and deep-frozen snow as if caught in a trap.

The Carthaginians make a road across a precipice.

tandem nēquīquam iūmentīs atque hominibus fatīgātīs,
castra in iugō posita, aegerrimē ad id ipsum locō pūrgātō: tantum
nivis fodiendum atque ēgerendum fuit. inde ad rūpem
35 mūniendam per quam ūnam via esse poterat militēs ductī, cum
caedendum esset saxum, arboribus circā immānibus dēiectīs
struem ingentem lignōrum faciunt eamque, cum et vīs ventī apta
faciendō ignī coorta esset, succendunt ārdentiaque saxa īnfūsō
acētō putrefaciunt. ita torridam incendiō rūpem ferrō pandunt
40 molliuntque ānfrāctibus modicīs clīvōs ut nōn iūmenta sōlum sed
elephantī etiam dēdūcī possent.

They reach the lowlands.

quadrīduum circā rūpem cōnsūmptum, iūmentīs prope fame
absūmptīs; nūda enim ferē;cacūmina sunt et, sī quid est pābulī,
obruunt nivēs. īnferiōra vallīs aprīcōsque quōsdam collēs habent
45 rīvōsque prope silvās et iam hūmānō cultū digniōra loca. ibi
iūmenta in pābulum missa et quiēs mūniendō fessīs hominibus
data. trīduō inde ad plānum dēscēnsum et iam locīs molliōribus
et accolārum ingeniīs.
hōc maximē modō in Italiam perventum est . . .

Monte Viso, which towers over the col by which Hannibal crossed the Alps

33   **posita (sunt).** **aegerrimē** with the greatest difficulty. **ad id ipsum** for this purpose. **pūrgātō: pūrgō** (1) I clear.

34   **fodiendum: fodiō** I hack out. **ēgerendum: ēgerō** I dig out.

35   **mūniendam: mūniō** I make passable. **ūnam** alone.

37   **struem lignōrum** pile of logs. **apta** fitted, suitable.

38   **succendunt** they kindle. **ārdentia** glowing.

38–39 **īnfūsō acētō** pouring on vinegar. **putrefaciunt** they cause to crumble. **torridam** heated. **pandunt** they make passable, make a road across.

40   **molliunt ānfrāctibus modicīs clīvōs** they make the slopes easier with short zig-zag tracks.

42   **quadrīduum . . . cōnsūmptum** four days were spent.

42–3 **prope absūmptīs** nearly killed.

43   **cacūmina** the peaks. **pābulī: pābulum** fodder.

44   **obruunt** cover. **nivēs** is the subject. **īnferiōra** the lower slopes (subject of **habent**). **vallīs = vallēs** (accusative plural). **aprīcōs quōsdem collēs** certain sunny hills = a number of sunny hills. of. **aprīcōs** sunny.

45   **cultū** habitation.

46   **mūniendō** by their efforts.

47–8 **trīduō** in three days time. **plānum** the plain. **molliōribus** goes with both **locīs** and **ingeniīs** (ablative absolute): they now went through places where both the terrain and the characters of the natives were milder.

48   **accolārum** of the inhabitants.

49   **maximē** by and large.

aliquot (indeclinable adjective)	several	**hauriō, haurīre, hausī, haustum**	I drain, drink up
circā (adverb or preposition + accusative)	around, round about	**immānis-e**	huge
		iūmentum, -ī, *n.*	pack animal
exitus, -ūs, *m.*	way out	**nūdus-a-um**	bare, unprotected, naked
genū, -ūs, *n.*	knee		
glaciēs, glaciēī, *f.*	ice	**vestīgium, vestīgiī,** *n.*	footprint, track
haereō, haerēre, haesī, haesum	I stick		

NB In this chapter you have met with many sentences in which the verb **esse** has been omitted. This happens frequently in Latin and normally the reader has no difficulty in supplying it.

Exercise 17.1

What does the prefix add to the meaning of each of the following words (all from this chapter)

prōlāpsa, subiectōs, prōcēdere, sustinēre, dēmittere, circumdūcere, īnfūsō, absūmptīs

Ⓖ Indicative and subjunctive in subordinate clauses

In Chapter 9 we saw that in main clauses the indicative is used to express facts, and the subjunctive is used for various types of non-factual expression (commands, wishes etc.).

The same principle applies to many types of subordinate clause:

1 Temporal clauses
puerī exspectāvērunt dum pater advēnit
The boys waited until their father arrived. (The **dum** clause expresses a fact.)
puerī exspectāvērunt dum pater advenīret
The boys waited until their father should arrive = waited for their father to arrive. (The subjunctive is used because the **dum** clause expresses a purpose, not a fact.)

2 Causal clauses
iūdicēs iuvenem condemnāvērunt quod senem occīdit
The jury condemned the young man because he had murdered the old man.
iūdicēs iuvenem condemnāvērunt quod senem occīdisset
The jury condemned the young man on the grounds that (because, as they thought,) he had murdered the old man.
(The subjunctive is used because what the **quod** clause expresses may not be a fact.)

3 Relative clauses
Caesar quīnque cohortēs relīquit quae nāvibus praesidiō erant
Caesar left behind five cohorts which guarded the ships.
Caesar quīnque cohortēs relīquit quae nāvibus praesidiō essent
Caesar left behind five cohorts which might guard the ships (= to guard the ships).
(The subjunctive is used because the relative clause expresses not a fact but a purpose in Caesar's mind.)

4 Concessive clauses
quamquam fessus es, festīnandum est
Although you are tired, you must hurry.
quamvīs fessus sīs, festīnandum est
However tired you may be, you must hurry.
(The subjunctive is used, because the **quamvīs** clause expresses not a fact, but a possibility.)

5 In indirect speech all subordinate clauses have their verbs in the subjunctive, because indirect speech expresses not facts but the words or thoughts of the speaker, which may or may not be true.

When Ambiorix attacked Sabinus's camp, he asked to have a parley with the Romans; he said to the men whom Sabinus sent to talk with him (see Chapter 4):
sē plūrimum Caesarī dēbēre, quod eius operā stīpendiō līberātus esset, quod Aduaticīs fīnitimīs suīs pendere cōnsuēvisset, quodque fīlius ab Caesare remissus esset, quem Aduaticī in servitūte tenuissent
(He said that) he owed a great deal to Caesar because through his efforts he had been freed from the tax he used to pay to his neighbours the Aduatici, and because his son had been sent back to him by Caesar, whom the Aduatici had held in slavery.

The verbs in the subordinate clauses – **quod . . . līberātus esset** (causal), **quod . . . cōnsuēvisset** (relative), **quod remissus esset** (causal), **quem . . . tenuissent** (relative) – are all in the subjunctive because they form part of what Ambiorix said.

Exercise 17.2

Translate
1 Brūtus Caesarem occīdere cōnstituit quod rēx fierī vellet.
2 negāvit eum quī rēgnum peteret dignum esse quī vīveret.
3 quamvīs ambitiōsus sīs, nōlī rēgnum tibi petere.
4 Brūtus amīcum quendam mīsit quī Caesarī persuādēret ut ad theātrum Pompēiī venīret.
5 in theātrō ānxiī exspectant dum Caesar perveniat.
6 multī dīxērunt Caesarem, postquam amīcum aggredientem vīdisset, 'et tū, Brūte' clāmāvisse.

Hannibal won a series of great victories over the Romans. The most devastating was the battle fought in 216 BC at Cannae, when perhaps 70,000 Romans were killed. However, though he stayed in Italy for sixteen years, he never succeeded in breaking the might of Rome. Eventually a Roman army led by Scipio took the war over to Africa and Hannibal was forced to return there. He was conclusively defeated by Scipio at the Battle of Zama in 202 BC.

Hannibal now set about rebuilding the prosperity of Carthage. But he had many enemies both among his fellow-citizens and at Rome, and seven years later he was exiled. We take up his story in 182 BC. King Prusias of Bithynia has guaranteed him a safe refuge but receives a distinguished Roman visitor.

Exercise 17.3

Translate the first paragraph and answer the questions on the rest

ad Prūsiam rēgem lēgātus T. Quīnctius Flāminīnus vēnit. ibi quia
ipse Prūsiās, ut grātificārētur praesentī Flāminīnō Rōmānīsque,
per sē necandī aut trādendī eius in potestātem cōnsilium cēpit, ā
prīmō colloquiō Flāminīnī mīlitēs extemplō ad domum
5   Hannibalis custōdiendam missī sunt.

    semper tālem exitum vītae suae Hannibal prōspexerat
animō, et Rōmānōrum inexpiābile odium in sē cernēns et fideī
rēgum nihil sānē cōnfīsus; Prūsiae vērō levitātem etiam expertus
erat; Flāminīnī quoque adventum velut fātālem sibi horruerat.
10   ad omnia undique īnfesta, ut iter semper aliquod praeparātum
fugae habēret, septem exitūs ē domō fēcerat, et ex iīs quōsdam
occultōs, nē custōdiā saepīrentur.

    sed tōtīus circuitum domūs ita custōdiīs complexī sunt, ut
nēmō inde ēlābī posset. Hannibal, postquam est nūntiātum
15   mīlitēs rēgiōs in vestibulō esse, postīcō, quod dēvium maximē
erat, fugere cōnātus, ut id quoque occursū mīlitum obsaeptum
sēnsit et omnia circā clausa custōdiīs dispositīs esse, venēnum,
quod multō ante praeparātum ad tālēs habēbat cāsūs, poposcit.
'līberēmus,' inquit, 'diuturnā cūrā populum Rōmānum, quandō
20   mortem senis exspectāre longum cēnsent. nec magnam nec
memorābilem ex inermī prōditōque Flāminīnus victōriam feret.'

    exsecrātus deinde in caput rēgnumque Prūsiae et hospitālēs
deōs violātae ab eō fideī testēs invocāns, pōculum exhausit. hic
vītae exitus fuit Hannibalis.

Prusias I of Bithyn[

Flamininus

2	**grātificārētur** (+ dative) he might do a favour to. **praesentī** who was on the spot.
3	**per sē** on his own initiative. **necandī** and **trādendī** are dependent on **cōnsilium**. **trādendī eius** of handing him over. **potestātem:** i.e. of the Romans.
3–4	**ā prīmō colloquiō** (+ gen.) as a result of his first conversation with ...
4	**extemplō** immediately.
6	**exitum vītae** end of life = death.
6–7	**prōspexerat animō** had foreseen. **inexpiāble** implacable. **in sē** against himself.
8	**nihil sānē** by no means. **vērō** indeed. **levitātem** untrustworthiness.
8–9	**expertus erat** he had had experience of.
9	**velut fātālem** as if (it would prove) fatal. **horruerat** he had felt fear at.
10	**ad omnia undique īnfesta** in the face of all the dangers which surrounded him. **iter** route.
10–11	**praeparātum fugae** prepared for flight = available for escape.
12	**nē custōdiā saepīrentur** so that they could not be blocked off by guards.
13	**circuitum** the perimeter. **custōdiīs complexī sunt** they (i.e. the king's men) surrounded with guards.

15	**in vestibulō** in the entrance-court. **posticō: posticum** back-door. **dēvium** secret.
16	**occursū mīlitum obsaeptum (esse)** was blocked off by the soldiers stationed there.
17	**circā** round about. **custōdiīs dispositīs** by the guards positioned there.
18	**praeparātum** ready. The story goes that he carried the poison with him concealed in a signet ring.
19	**diuturnā cūrā** from (their) long-lasting anxiety.
19–20	**quandō mortem senis exspectāre longum cēnsent** since they consider it (a) long (business) to wait for the death of an old man, i.e. they are too impatient to wait for me to die in the course of nature.
22	**exsecrātus in** (+ accusative) having cursed. **caput** the life.
22–3	**hospitālēs deōs** the gods of hospitality. **violātae** violated, abused (agreeing with **fideī**). **testēs invocāns** calling upon as witnesses of ... **pōculum** the draught, drink.

1 What two factors have made Hannibal foresee his death (ll.6–9)?

2 What precautions has he taken?

3 Why cannot he make his getaway?

4 How does he kill himself ? What attitude to the Romans does he express in his final speech? How admirable do you find him in his death?

5 How does the character of King Prusias come across to you in this passage?

6 Hannibal dies cursing Prusias. What is the nature of his curse? In what way does this re-echo Dido's curse on Aeneas? How dramatically appropriate do you find it?

7 Distinguish the uses of the word **ut** in lines 10, 13 and 16.

8 Explain the subjunctive **līberēmus** in l.19.

Exercise 17.4

Translate into Latin

1 The snow is so deep that I can scarcely walk.

2 Hannibal showed his men Italy in order to encourage them.

3 When we have gone down into Italy, the Romans will not be able to resist us.

4 If Hannibal had despaired, the whole army would have been destroyed.

5 Hannibal asked his men whether they wished to rest or to set out.

Exercise 17.5

Translate into Latin

Once a show of wild animals was seen at Rome. All the animals were savage, but one lion was especially large. A slave of the consul was led into the arena (**arēna, -ae,** *f.*) to fight with it. The whole crowd became silent because they wondered whether the slave would turn and run or would fight with the lion. The lion, which was approaching the man, suddenly stopped and then embraced him and licked (**lambō, -ere, lambī**) him with his tongue. All were so astonished that they shouted with joy.

Abbreviations of names

Learn the following abbreviations of common Roman names

A.	Aulus	M.	Mārcus	D.	Decimus	T.	Titus
C.	Gāius	P.	Pūblius	L.	Lucius	Ti.	Tiberius
Cn.	Gnaeus	Q.	Quīntus	Sex.	Sextus		

The satirical writer Juvenal, who wrote in the first half of the second century AD, used the story of Hannibal to demonstrate what he saw as the ultimate futility of military ambition. After you have read this passage, you might consider what your attitude is to the achievements of Hannibal or any other great military leader.

A balance from Pompeii

> expende Hannibalem: quot lībrās in duce summō
> inveniēs? hic est, quem nōn capit Africa Maurō
> percussa ōceanō Nīlōque admōta tepentī.
> additur imperiīs Hispānia, Pȳrēnaeum
> 5 trānsilit. opposuit nātūra Alpemque nivemque:
> dīdūcit scopulōs et montem rumpit acētō.
> iam tenet Italiam, tamen ultrā pergere tendit.
> 'āctum,' inquit, 'nīl est, nisi Poenō mīlite portās
> frangimus et mediā vexillum pōnō Subūrā.'
> 10 ō quālis faciēs et quālī digna tabellā,
> cum Gaetūla ducem portāret bēlua luscum.
> exitus ergō quis est? ō glōria, vincitur īdem
> nempe et in exsilium praeceps fugit atque ibi magnus
> mīrandusque cliēns sedet ad praetōria rēgis
> 15 dōnec Bīthȳnō libeat vigilāre tyrannō.
> fīnem animae quae rēs hūmānās miscuit ōlim,
> nōn gladiī, nōn saxa dabunt nec tēla, sed ille
> Cannārum vindex et tantī sanguinis ultor
> ānulus. ī dēmēns et saevās curre per Alpēs
> 20 ut puerīs placeās et dēclāmātiō fīās. Juvenal, *Satires*, X, 147–67

1	**expende** weigh in the scale. **quot lībrās** (*acc. pl.*) how many pounds (of weight), i.e. how much do the ashes of the dead Hannibal weigh?
2	**nōn capit** cannot contain.
2–3	**Maurō/percussa ōceanō** pounded by the Moorish Ocean (i.e. the Atlantic). **Nīlō admōta tepentī** stretching as far as the warm Nile.
4	**Pȳrēnaeum** the Pyrenees.
5	**trānsilit** he leaps over. **opposuit** put in his way. **Alpemque nivemque** the Alpine snows (hendiadys: literally, 'the Alp and the snow').
6	**dīdūcit scopulōs** he splits the rocks. **acētō** with vinegar.
7	**pergere** to proceed. **tendit** he strives.
8	**Poenō mīlite:** singular for plural, i.e. 'with the Carthaginian soldiers'. **portās** the gates (of Rome).
9	**vexillum** standard. **Subūrā:** a district of Rome. This is humorous since it was the red-light district.
10	**faciēs** sight. **quam digna tabellā** how worthy of a picture! = how fit for caricature!
11	**Gaetūla bēlua** an African monster, i.e. an elephant. **luscum** one-eyed (agreeing with **ducem**). In 217 B.C. Hannibal crossed the Apennines on his one surviving elephant and lost an eye through disease.
12	**exitus** his end. **vincitur:** Hannibal suffered his one and only defeat at Zama in 202 B.C. **īdem** that same man.
13	**nempe:** the word conveys scorn.
14	**cliēns** client, hanger-on. **praetōria** palace.
15	**dōnec** until. **libeat** (+ dative) it might please. **vigilāre** to wake up (and admit his visitor). **tyrannō:** the Greek word for 'king'.
16	**animae** life, spirit: dative, indirect object of **dabunt**. **miscuit** turned upside-down.
18	**Cannārum vindex** avenger of Cannae (216 B.C., Hannibal's greatest victory over the Romans). **ultor** (+ genitive) exacting revenge for.
19	**ānulus** a ring (in which he carried his poison). **dēmēns** madman.
20	**dēclāmātiō** a subject for debates in schools.

Gaetūla bēlua

CHAPTER XVIII

Ovid tells the story of his life

Ovid was born on 20 March 43 BC in Sulmo (modern Sulmona) in the Apennines about ninety miles east of Rome. He had one brother, born exactly a year before. Their family was an old-established and fairly wealthy equestrian one and Ovid had the right background to pursue a successful senatorial career. But he found himself irresistibly drawn to poetry. He was to be the last of the Augustan love poets.

Sulmo

> *Ovid introduces himself and his brother.*
> ille ego quī fuerim, tenerōrum lūsor amōrum,
>   quem legis, ut nōrīs, accipe posteritās.
> Sulmo mihī patria est, gelidīs ūberrimus undīs,
>   mīlia quī noviēs distat ab urbe decem.
> 5   ēditus hīc ego sum, nec nōn, ut tempora nōrīs,
>   cum cecidit fātō cōnsul uterque parī.
> nec stirps prīma fuī; genitō sum fratre creātus,
>   quī tribus ante quater mēnsibus ortus erat.
> Lūcifer ambōrum nātālibus affuit īdem:
> 10   ūna celebrāta est per duo lība diēs.

The statue of Ovid at Sulmo

1–2   *accipe* (listen), *posteritās* (future generations), *ut nōrīs (=nōverīs) quī (=quis) ille ego fuerim, lūsor tenerōrum amōrum quem legis.*
    ille ego that famous I, i.e. I, the famous poet Ovid (**ille** frequently means 'the famous').   **lūsor** playful poet.

3   **gelidīs . . . undīs** with ice-cold waters.

4   **noviēs decem** nine times ten = ninety.   **distat** is away.   **urbe**: i.e. Rome.

5   **ēditus** born.   **nec nōn** and also.   **nōrīs = nōverīs.**

6   **fātō . . . parī** by the same fate. The two consuls Hirtius and Pansa were killed fighting Antony in the year of Ovid's birth (see Part II, Chapter 7).

7   **stirps prīma** the first born.   **genitō . . . fratre** my brother already born = after the birth of my brother.

8   **tribus . . . quater** four times three, i.e. twelve.

9   **Lūcifer . . . īdem** the same dawn.   **nātālibus** at the birthday: Ovid and his brother had the same birthday.   **affuit = adfuit.**

10   **ūna . . . diēs** one day (**diēs**, referring to a specific day, can be feminine).   **lība,** *n.pl.* cakes: offered to the deity who protected the birthday child.

The education of the poet and his brother.
prōtinus excolimur tenerī cūrāque parentis
    īmus ad īnsignēs urbis ab arte virōs.
frāter ad ēloquium viridī tendēbat ab aevō
    fortia verbōsī nātus ad arma forī;
15   at mihi iam puerō caelestia sacra placēbant,
    inque suum fūrtim Mūsa trahēbat opus.

The nine Muses

saepe pater dīxit, 'studium quid inūtile temptās?
    Maeonidēs nūllās ipse relīquit opēs.'
mōtus eram dictīs, tōtōque Helicōne relictō
20   scrībere temptābam verba solūta modīs.
sponte suā carmen numerōs veniēbat ad aptōs,
    et quod temptābam scrībere versus erat.

They both prepare for a career in public life.
intereā tacitō passū lābentibus annīs
    līberior frātrī sūmpta mihīque toga est,
25   induiturque umerīs cum lātō purpura clāvō,
    et studium nōbīs, quod fuit ante, manet.
iamque decem vītae frāter gemināverat annōs,
    cum perit, et coepī parte carēre meī.
cēpimus et tenerae prīmōs aetātis honōrēs,
30   ēque virīs quondam pars tribus ūna fuī.

192

11   **prōtinus excolimur tenerī** we started our education right away, while still young.

12   *ad virōs urbis ab arte īnsignēs:* **īnsignēs ab arte** famous because of their skill. Like Horace, Ovid went to Rome to be educated.

13   **ad ēloquium** on public speaking.   **viridī . . . ab aevō** from his youngest years, literally, 'from his green age'.   **tendēbat** was keen.

14   **ad arma** for the warfare.   **verbōsī . . . forī** of the wordy law court. The lawcourts were in the Forum.

15   **caelestia sacra** the holy rites of poetry (inspiration comes from heaven (**caelum**)).

16   **trahēbat:** understand **mē.**

A young man in a toga

18   **Maeonidēs:** i.e. Homer, from Maeonia (=Lydia).

19   **Helicōne** (ablative): Mount Helicon in Boeotia was the home of the Muses. Ovid means that he totally abandoned poetry.

20   **verba solūta modīs** words freed from metre, i.e. prose.

21   **numerōs . . . ad aptōs** into the appropriate metre.

24   **līberior . . . toga** the toga which gave more freedom, i.e. the adult's toga, the **toga virīlis,** which was worn (**sūmpta . . . est**) by boys from about the age of sixteen.   **frātrī . . . mihīque** by my brother and me (dative of the agent).

25   **induitur** was put on.   **cum lātō purpura clāvō** the purple (toga) with the broad stripe = the toga with the broad purple stripe. The **lātus clāvus** was worn by senators, sons of senators and **equitēs illūstrēs**, i.e. sons of knights whom Augustus wished to encourage to embark on a senatorial career.

26   Ovid's brother remained an enthusiast for the law, Ovid for poetry.

27   **gemināverat** had doubled; i.e. he had reached the age of twenty.

29–30   *et cēpimus* and I took. **tenerae prīmōs aetātis honōrēs** the first offices open to (someone of my) tender age.

30   Ovid served as one of a Board of Three, dealing with prisons or the mint.

But Ovid abandons this ambition in favour of poetry.
cūria restābat: clāvī mēnsūra coācta est;
    māius erat nostrīs vīribus illud onus.
nec patiēns corpus, nec mēns fuit apta labōrī,
    sollicitaeque fugāx ambitiōnis eram,
35   et petere Aoniae suādēbant tūta sorōrēs
    ōtia, iūdiciō semper amāta meō.

His admiration for older poets of his time.
temporis illīus coluī fōvīque poētās
    quotque aderant vātēs, rēbar adesse deōs.
saepe suōs solitus recitāre Propertius ignēs,
40   iūre sodāliciī, quō mihi iūnctus erat.
et tenuit nostrās numerōsus Horātius aurēs,
    dum ferit Ausoniā carmina culta lyrā.
Vergilium vīdī tantum: nec avāra Tibullō
    tempus amīcitiae fāta dedēre meae.

Horace

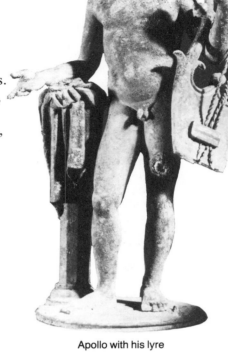

Apollo with his lyre

He publishes his first poems which are inspired by Corinna.
45   utque ego māiōrēs, sīc mē coluēre minōrēs,
    nōtaque nōn tardē facta Thalīa mea est.
carmina cum prīmum populō iuvenālia lēgī,
    barba resecta mihī bisve semelve fuit.
mōverat ingenium tōtam cantāta per urbem
50   nōmine nōn vērō dicta Corinna mihi.
molle Cupīdineīs nec inexpugnābile tēlīs
    cor mihi, quodque levis causa movēret, erat.
cum tamen hic essem minimōque accenderer ignī,
    nōmine sub nostrō fābula nūlla fuit.

Ovid falls foul of the Emperor.
55   iam mihi cānitiēs pulsīs meliōribus annīs
    vēnerat, antīquās miscueratque comās,
cum maris Euxīnī positōs ad laeva Tomītās
    quaerere mē laesī prīncipis īra iubet.

194

31	**restābat** awaited (me). At the age of thirty, he could start on the **cursus honōrum** by becoming a quaestor. **clāvī mēnsūra coācta est** the width of my purple stripe was narrowed. He withdrew from public life, becoming an **eques** of the second class and thus having a narrower stripe on his toga.
32	**māius** too great for . . .
33	**nec patiēns (labōris) corpus (erat).**
34	**sollicitae fugāx ambitiōnis** a runaway from political life with all its anxieties: literally **ambitiō** means 'going round', canvassing for a political cause.
35–6	**Aoniae sorōrēs** the sisters from Mount Helicon in Boeotia (=Aonia), i.e. 'the Muses'. *Aoniae sorōrēs suādēbant (mihi) petere tūta ōtia.*
38	**quot aderant vātēs** as many poets as were present = any poets who were present.
39	**solitus (est). suōs ignēs** his fiery poetry = his love poetry.
40	**iūre sodāliciī** by the ties of comradeship.
41	**numerōsus** with his many metres.
42	**ferit Ausoniā carmina culta lyrā** accompanied his sophisticated poetry on his Ausonian (=Italian) lyre: **ferit** struck, i.e. struck the strings with his plectrum.
43	**tantum** only; i.e. Ovid never spoke to him. Tibullus died when Ovid was about twenty four.
43–4	**avāra . . . fāta** greedy fate. **dedēre = dedērunt.** Fate gave Tibullus no time for friendship with Ovid (**meae amīcitiae**).
45	**māiōrēs . . . minōrēs** older (poets) . . . younger (poets).
46	**Thalīa mea** my Muse. Thalia was the Muse of comedy and light verse.
47	**lēgī** I read aloud. Poetry was usually first made known by a public recitation.
48	**barba . . . mihī** my beard. **resecta (erat)** had been cut. **bisve semelve** once or twice, i.e. when he was eighteen or so.
49–50	**mōverat ingenium** had stirred my genius: **Corinna** is the subject of **mōverat. dicta . . . mihī** called by me. It was customary for poets to give their mistress a 'poetical' name (**nōmine nōn vērō**); cf. Catullus's Lesbia. Corinna, Ovid's first love (if she was a real person), is the subject of many of his early love poems.
51	**Cupīdineīs nec inexpugnābile tēlīs** and not proof against the darts of Cupid: take **nec** before **Cupīdineīs.**
52	**quod** + subjunctive: generic, i.e. 'the sort of heart which . . .'.
53	**hic** like this. **accenderer** could be set on fire.
54	**fābula** scandal. **sub** attached to.
55	**cānitiēs** white hair, old age. **pulsīs meliōribus annīs** when my better (= more youthful) years were gone.
56	**miscuerat** had flecked, had sprinkled.
57–8	*cum īra prīncipis mē iubet quaerere* (= to make for) *Tomītās:* **maris . . . Euxīnī positōs ad laeva Tomītās** the people of Tomis who live (literally, are placed) on the left of the Euxine Sea (i.e. the Black Sea). **laesī prīncipis** of the offended emperor. Augustus banished Ovid in AD 8 when the poet was fifty-one. We shall discuss the reasons for this banishment in the final chapter.

Ovid's last night in Rome.

cum subit illīus trīstissima noctis imāgō,

60   quae mihi suprēmum tempus in urbe fuit,

cum repetō noctem, quā tot mihi cāra relīquī,

        lābitur ex oculīs nunc quoque gutta meīs.

iam prope lūx aderat, quā mē discēdere Caesar

        fīnibus extrēmae iusserat Ausoniae.

65   nec spatium nec mēns fuerat satis apta parandī:

        torpuerant longā pectora nostra morā.

nōn mihi servōrum, comitēs nōn cūra legendī,

        nōn aptae profugō vestis opisve fuit.

nōn aliter stupuī, quam quī Iovis ignibus ictus

70   vīvit et est vītae nescius ipse suae.

He bids farewell to his friends and his wife.

ut tamen hanc animī nūbem dolor ipse remōvit,

        et tandem sēnsūs convaluēre meī,

adloquor extrēmum maestōs abitūrus amīcōs,

        quī modo dē multīs ūnus et alter erant.

75   uxor amāns flentem flēns ācrius ipsa tenēbat,

        imbre per indignās usque cadente genās.

nāta procul Libycīs aberat dīversa sub ōrīs,

        nec poterat fātī certior esse meī.

quōcumque aspicerēs, luctūs gemitūsque sonābant,

80   formaque nōn tacitī fūneris intus erat.

His wife wishes to follow him into exile.

tum vērō coniūnx umerīs abeuntis inhaerēns

        miscuit haec lacrimīs trīstia verba meīs:

'nōn potes āvellī. simul hinc, simul ībimus,' inquit;

        'tē sequar et coniūnx exulis exul erō.

85   tē iubet ē patriā discēdere Caesaris īra,

        mē pietās. pietās haec mihi Caesar erit.'

tālia temptābat, sīcut temptāverat ante,

        vixque dedit victās ūtilitāte manūs.

Caesar Augustus

196

59	**subit** comes to my mind. **imāgō** the picture.		

59   **subit** comes to my mind.   **imāgō** the picture.

60   **quae**: the antecedent is **noctis.**   **suprēmum** last, final.

61   **repetō** I recall.

62   **gutta** a teardrop.

63   **prope** nearly.   **lūx** the dawn, i.e. the day.

66   **fīnibus ... extrēmae Ausoniae**: from the boundaries of furthest Italy = from the furthest boundaries of Italy.

65   **spatium** time.   **nec spatium nec mēns ... parandī** neither time nor spirit for getting ready.   **satis apta** sufficiently suitable things = the sort of things which were needed.

66   **torpuerant** had become numb.

67–8   *nōn (fuit cūra) mihi servōrum (legendōrum), nōn (mihi fuit) cūra legendī comitēs*. **profugō** for an exile.   **vestis opisve** genitives; understand **cūra** 'I took no thought for ...'.   **opis** what could be of help to me.

69   **nōn aliter ... quam** not otherwise than = just like.   **stupuī** I was dumbfounded.   **Iovis ignibus ictus** struck by the thunderbolt (literally, by the fires) of Jupiter.

70   **et** (here) = though.   **nescius** (+ genitive) unconscious of.

72   **sēnsūs convaluēre** (=**convaluērunt**) my senses revived.

73   **extrēmum** for the last time.

74   **modo dē multīs** out of the many (I had had) recently.   **ūnus et alter** one or two.

75   **ācrius**: his wife is in even greater anguish than he is.

76   **imbre** (ablative) a shower (of tears).   **per indignās genās** down her cheeks which did not deserve (such disfigurement).   **usque** continually.

77   **nāta** (my) daughter.   **dīversa** distant.   **sub** on, near.

78   **fātī certior esse meī** = **certior fierī dē fātō meō.**

79   **quōcumque aspicerēs** wherever one (literally, you) might look.   **luctūs gemitūsque sonābant** there was the noise of mourning and groaning.

80   **forma nōn tacitī fūneris** the appearance of a not quiet (= noisy) funeral.   **intus** inside (the house).

81   **umerīs abeuntis inhaerēns** clinging to the shoulders of (me) going away = with her arms flung tightly round me as we parted.   **miscuit** mingled.

83   **nōn potes āvellī** you cannot be torn away = I shall not allow you to be torn from me.

84   **exul, exulis,** *c.* an exile.

86   **mē (iubet) pietās.**

87   **sīcut** just as.

88   **dedit victās ... manūs** literally, 'gave her hands, conquered by ...'   **manūs dō** is an expression meaning 'I surrender'. Translate 'she surrendered, won over by ...'   **ūtilitāte** by practical considerations. She stayed in Rome to work for Ovid's recall.

aptus-a-um	fit, suitable	occidō, occidere, occidī	I die, fall
coma, comae, *f.*	hair	sonus, -ī, *m.*	sound
cor, cordis, *n.*	heart	sponte (meā, tuā, suā etc.)	of (one's) own
feriō, ferīre	I beat, strike		accord
fūrtim	secretly, stealthily	tener, tenera, tenerum	tender, young
gemitus, -ūs, *m.*	groaning	umerus, -ī, *m.*	shoulder
maestus-a-um	sad		

G Word building

Make sure that you know the following compounds of
gradior, gradī, gressus

aggredior, aggredī,	I attack	**ēgredior**	I go out, disembark
aggressus		**ingredior**	I go in
congredior	I meet	**prōgredior**	I go forward, advance
dīgredior	I depart	**regredior**	I go back

Give English derivations from as many of these compounds as you can.

G 'cum'

1 **cum** = when
The general rule, which you learnt in Part II, is that **cum** = 'when'
takes the indicative in present and future time and the subjunctive in
past time. But you will meet **cum** followed by the indicative even in
past time under various special circumstances, e.g. if **cum** = 'precisely
when', if **cum** = 'whenever', if the idea of time is in the main clause
and not in the **cum** clause. Study the following examples of **cum** + the
indicative.

cum haec legēs, Rōmae erō
When you read this, I shall be in Rome.
cum Rōmam advēnerō, ab Orbiliō docēbor
When I get to Rome, I shall be taught by Orbilius.

NB Though we translate the verbs in the **cum** clauses above as if they
were present tenses in English, they in fact refer to the future and so
will be future or future perfect in Latin. 'I shall tell you when I see
you' means 'I shall tell you when I shall see you' or 'I shall tell you
when I shall have seen you.'

sōl oriēbātur cum ad lūdum advēnī
The sun was rising when I got to the school.
cum Caesar advēnit, Gallī rebelliōnem fēcerunt
When Caesar arrived, the Gauls rebelled.
cum Horātiī carmina audīverat, Ovidius ea valdē admīrātus est
Whenever he heard Horace's poems, Ovid admired them very much.

NB When **cum** = 'whenever', it is followed by the pluperfect
indicative.

2 Remember that **cum** + the subjunctive can mean 'since' or
'although'. The context will usually make it clear which meaning is
intended.
quae cum ita essent, laetissimus eram
Since these things were so, I was very happy.
cum in fundō meō laetissimus sim, ad urbem tamen redībō
Although I am very happy on my farm, I shall still return to the city.

198

Exercise 18.1

Translate

1 cum ad urbem advēnerō, ad hominēs doctōs ībō.
2 cum Vergilius recitāverat, admīrābar.
3 vesper iam aderat cum Vergilius fīnem recitandī fēcit.
4 cum subit illius noctis imāgō, lacrima ex oculīs nunc quoque lābitur.
5 cum pater meus mē senātōrem fierī velit, ego carmina scrībam.
6 iam lūx aderat cum Italia fuit relinquenda.
7 Ovidius laetior fiēbat cum uxor eum comitārī cuperet.

'dum'

dum (= while) is always followed by the present indicative, except when the actions of the verbs in the **dum** clause and the main clause begin and end at the same time (i.e. when **dum** means 'as long as'):

dum Caesar abest, Gallī rebelliōnem fēcerunt
While Caesar was away, the Gauls rebelled.

lūdēbat dum licēbat
He played while (= as long as) he was allowed to.

Exercise 18.2

Translate

1 dum in urbe sum, ēloquium didicī.
2 dum Horātius recitābat, omnēs intente audiēbant.
3 dum haec dīcō, Corinna exiit.
4 dum haec Rōmae aguntur, Gallī mīlites convocāvērunt.

Exercise 18.3

Translate into Latin

1 When I arrive at Ovidius's house, I shall see Corinna.
2 Corinna went out while Ovidius was reciting.
3 The sun was setting when she returned.
4 Since she was so beautiful, I forgave her.
5 When Horatius was reciting yesterday, she listened eagerly.
6 Will she listen to me when I recite my poetry?

quōcumque aspicerēs, luctūs gemitūsque sonābant

Exercise 18.3

Read the following passage and then answer the questions after it

The death of Corinna's parrot

psittacus, Eōīs imitātrix āles ab Indīs,
  occidit: exsequiās īte frequenter, avēs;
īte, piae volucrēs, et plangite pectora pinnīs
  et rigidō tenerās ungue notāte genās.
5 omnēs, quae liquidō lībrātis in āere cursūs,
  tū tamen ante aliōs, turtur amīce, dolē.
plēna fuit vōbīs omnī concordia vītā
  et stetit ad fīnem longa tenāxque fidēs.
quid tamen ista fidēs, quid rārī forma colōris,
10   quid vōx mūtandīs ingeniōsa sonīs,
quid iuvat, ut datus es, nostrae placuisse puellae?
  īnfēlīx avium glōria nempe iacēs.
occidit illa loquāx hūmānae vōcis imāgō
  psittacus, extrēmō mūnus ab orbe datum.
15 septima lūx vēnit nōn exhibitūra sequentem;
  clāmāvit moriēns lingua 'Corinna, valē.'

1	**psittacus** parrot. **Eōīs imitātrix āles ab Indīs** imitating bird from India (literally, the Indi) in the East.
2	**occidit**:note scansion – not **occīdit**. **exsequiās** to the funeral. **frequenter** in flocks.
3	**volucrēs** winged creatures. **plangite ... pinnīs** beat with your wings.
4	**rigidō ... ungue** with your stiff claws (singular for plural). **notāte genās** mark (= tear) your cheeks. At Roman funerals, the mourners beat their breasts and tore their cheeks.
5	**omnēs**: understand **dolēte**. **quae liquidō lībrātis in āere cursūs** who wing your way (literally, balance your journeys) through the clear air.
6	**ante aliōs** before the others = more than all the others. **turtur** turtle dove.
7	**omnī** agrees with **vītā**. **concordia** harmony.
8	**stetit ad fīnem** lasted to the end. **tenāx** firm.
9–10	**quid ... quid ... quid ...** understand **iuvat** (1.11) what help is ...? **ingeniōsa** skilled in .... Now the parrot is being addressed.
11	**quid iuvat ... placuisse** what help is it to have pleased = what good has it done you that you gave pleasure to. **ut** as soon as.
12	Ovid answers the questions he has just asked: **quid iuvat ...**? All was of no avail, for the glory of the bird world lies dead. **nempe** (= to be sure, but perhaps better left untranslated) confirms this answer.
13	**loquāx ... imāgō** garrulous echo.
14	**extrēmō mūnus ab orbe** a present from the edge of the world.
15	**lūx** dawn, i.e. the seventh day after the onset of his illness. **nōn exhibitūra sequentem** not about to reveal a following (dawn), i.e. the bird died on the seventh day.

1 Where did the parrot come from? Who do you suppose gave it to Corinna?
2 Why is the turtle dove invited to grieve especially?
3 **quid** ... **quid** ... **quid** ... **quid** ...? (ll.9–11) – what is the answer to these questions? (one word) What qualities of the bird does Ovid list here and which does he intend to be the most important?
4 Translate ll.13–14.
5 Scan l.7. Which adjectives agree with which nouns?
6 In l.3 how many words start with 'p'? How does the sound of this line help to express the sense?
7 In l.4 why does Ovid put the adjectives **rigidō** and **tenerās** next to each other?
8 Write a brief character sketch of Corinna's parrot.
9 How are the birds in the poem given human characteristics? Quote three examples of where this is done.
10 How seriously do you think Ovid wants this poem to be taken?
11 If you have read Catullus's poem about a sparrow on p.112, the differences and similarities between the two poems.

turturēs

Exercise 18.4

Translate into English

The emperor Nero began to hate his mother so much that he decided to kill her, and he thought how he might most easily do such a deed. At length an evil friend advised him with these words: 'Build a ship which will collapse while your mother is on it.' Nero gave his mother dinner and then led her to the ship. Although the sea was calm, the ship collapsed. But Agrippina jumped into the sea and swam to the shore. If she had not done this, she would certainly have died.

CHAPTER XIX

Ovid the lover

Ovid called himself 'tenerōrum lūsor amōrum' (the poet who writes playfully of tender love). But his love poetry is not only tender and playful. On occasion it can be very sensuous.

1 *Siesta time*
The scene is set.

> Aestus erat, mediamque diēs exēgerat hōram;
>   adposuī mediō membra levanda torō.
> pars adaperta fuit, pars altera clausa fenestrae,
>   quāle ferē silvae lūmen habēre solent,
> 5 quālia sublūcent fugiente crepuscula Phoebō
>   aut ubi nox abiit nec tamen orta diēs.
> illa verēcundīs lūx est praebenda puellīs,
>   quā timidus latebrās spēret habēre pudor.
>
> *Corinna's arrival – and the sequel.*
> ecce, Corinna venit tunicā vēlāta recinctā,
> 10   candida dīviduā colla tegente comā,
> dēripuī tunicam; nec multum rāra nocēbat,
>   pugnābat tunicā sed tamen illa tegī;
> cumque ita pugnāret tamquam quae vincere nōllet,
>   victa est nōn aegrē prōditiōne suā.
> 15 ut stetit ante oculōs positō vēlāmine nostrōs,
>   in tōtō nusquam corpore menda fuit:
> quōs umerōs, quālēs vīdī tetigīque lacertōs!
>   fōrma papillārum quam fuit apta premī!
> quam castīgātō plānus sub pectore venter!
> 20   quantum et quāle latus! quam iuvenāle femur!
> singula quid referam? nīl nōn laudābile vīdī,
>   et nūdam pressī corpus ad usque meum.
> cētera quis nescit? lassī requiēvimus ambō.
>   prōveniant mediī sīc mihi saepe diēs!

in tōtō nusquam corpore menda fuit

1	**aestus erat** there was heat = it was sultry. **exēgerat** had passed.
2	**adposuī** I placed, lay. **mediō . . . torō** on the middle of the bed, i.e. he was alone. **levanda** to be rested, to rest.
3	**pars (fenestrae) adaperta (= aperta) fuit (fenestra** window).
4	**quāle . . . lūmen** the sort of light which . . . **ferē** generally.
5	**quālia . . . crepuscula** the sort of twilight which . . . (poetic plural). **sublūcent** glimmers. **Phoebō**: Phoebus, god of the sun.
7	**verēcundīs . . . puellis** for shy girls.
8	**latebrās spēret habēre** may hope to have hiding places = may hope to hide.
9	**tunicā vēlāta recinctā** dressed in a loose tunic.
10	**colla**, *n.pl.* neck. **dīviduā** (ablative) parted.
11	**dēripuī** I snatched off. **nec multum rāra nocēbat** (since it was) thin (**rāra**), it did not do much harm, i.e. it did not really hide anything.
12	translate **sed tamen** at the start of the line. **pugnābat . . . tegī** she struggled to be covered. The infinitive expresses purpose. She puts on a show of modesty.
13	**tamquam quae** (+ subjunctive) like a woman who . . .
14	**nōn aegrē** not with difficulty = easily. **prōditiōne suā** by her self-betrayal.
15	**positō . . . vēlāmine** her clothing put aside.
16	**menda** blemish.
17	**lacertōs** arms.
18	**papillārum** of her breasts.
19	**castīgātō** disciplined = well-formed. **plānus** flat.
20	**quam iuvenāle femur** how youthful her thigh.
21	**singula quid referam?** why should I mention each (of her charms) individually? **nīl nōn laudābile** nothing which was not worthy of praise.
22	**(eam) nūdam. corpus ad usque meum** tightly to my body.
23	*quis nescit cētera.* **lassī** exhausted.
24	**prōveniant** may they turn out (the subjunctive expresses a wish).

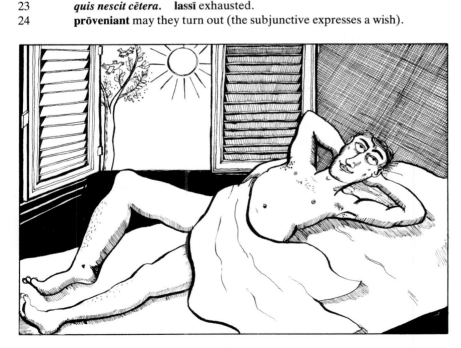

adposuī membra
levanda torō

Siesta time is from Ovid's collection of love poetry called the <u>Amōrēs.</u>
He began this in about 25 BC when he was not yet twenty. A quarter of a
century later, he wrote the <u>Ars Amātōria</u> (<u>The Art of Love</u>), a manual
on the technique of seduction. The immorality of this poem was one of
the causes of Ovid's exile.
Here is an excerpt from it.

2 *The theatre is an excellent place to pick up girls.*

> sed tū praecipuē curvīs vēnāre theātrīs;
>> haec loca sunt vōtō fertiliōra tuō.
> illīc inveniēs quod amēs, quod lūdere possīs,
>> quodque semel tangās, quodque tenēre velīs.

5
> ut redit itque frequēns longum formīca per agmen,
>> grāniferō solitum cum vehit ōre cibum,
> aut ut apēs saltūsque suōs et olentia nactae
>> pāscua per flōrēs et thyma summa volant,
> sīc ruit ad celebrēs cultissima fēmina lūdōs;

10
>> cōpia iūdicium saepe morāta meum est.
> spectātum veniunt, veniunt spectentur ut ipsae;
>> ille locus castī damna pudōris habet.

curvīs vēnāre theātrīs

> *Romulus started this Roman tradition.*
> prīmus sollicitōs fēcistī, Rōmule, lūdōs,
>> cum iūvit viduōs rapta Sabīna virōs.

15
> in gradibus sēdit populus dē caespite factīs,
>> quālibet hirsūtās fronde tegente comās.
> respiciunt oculīsque notant sibi quisque puellam
>> quam velit, et tacitō pectore multa movent.

1 **tū**: i.e. the young man whom Ovid is advising. **praecipuē** especially. **curvīs** curving. This refers to the curving tiers of the theatre. **vēnāre**: imperative of **vēnor** I hunt.

2 **loca**: the plural of **locus** can be neuter. **vōtō fertiliōra tuō** more productive than your wish = even more productive than you could wish.

3 **quod amēs** (something) which you may love = a girl to love. **lūdere** to deceive.

4 **semel tangās** you may touch once (and then leave) = have a brief affair with.

5 **ut** as. **redit itque** hurries to and fro. **frequēns . . . formīca** many an ant. **per agmen** along the column = in a long column.

6 *cum vehit solitum cibum grāniferō ōre.* **solitum** usual. **grāniferō** grain-carrying.

7 **apēs** bees. **saltūs suōs et olentia nactae/pāscua** having got to (literally, having obtained) their meadows and fragrant pastures = haunting the meadows, etc. **per** over. **summa thyma** the top of the thyme.

9 **cultissima fēmina** the most fashionable ladies (singular for plural).

10 **morāta est** has delayed = has confused. He can't make up his mind because of the *embarras de richesses*.

11 **spectātum** to see. The supine expresses purpose after a verb of motion. *veniunt ut ipsae spectentur*.

12 **castī damna pudōris** the losses of chaste modesty = the destruction of chaste modesty.

13 **sollicitōs** disturbed, chaotic. In its early years, Rome desperately needed women to keep its population going. Romulus invited the neighbouring Sabines to a show at Rome, planning to kidnap the women.

14 *rapta Sabīna iūvit viduōs virōs:* the stolen Sabine woman pleased the wifeless men = the wifeless men thought it a good idea to steal the Sabine women (**rapta** (from **rapiō**) snatched, stolen).

15 **dē caespite** from turf.

16 **quālibet . . . fronde** (ablative) any old foliage. They break off branches to act as sunshades (the rape occurred in August). **hirsūtās** shaggy.

17 **respiciunt** look back at. In Augustan times, the women sat in the back rows of the theatres. Ovid makes this the custom in early Rome as well. **notant** mark out: plural because **quisque** (=each one) refers to all of them.

18 **multa movent** they move many things = they feel deep emotion.

Romulus gives the sign to seize the women.
in mediō plausū (plausūs tunc arte carēbant)
20   rēx populō praedae signa petīta dedit.
prōtinus exsiliunt, animum clāmōre fatentēs
     virginibus cupidās īniciuntque manūs;
ut fugiunt aquilās, timidissima turba, columbae
     utque fugit vīsōs agna novella lupōs,
25   sīc illae timuēre virōs sine lēge ruentēs;
     cōnstitit in nūllā quī fuit ante color.

virginibus īniciunt manūs

Hysteria takes the women in many different ways.
nam timor ūnus erat, faciēs nōn ūna timōris:
     pars laniat crīnēs, pars sine mente sedet;
altera maesta silet, frūstrā vocat altera mātrem;
30   haec queritur, stupet haec; haec manet, illa fugit.
sī qua repugnārat nimium comitemque negārat,
     sublātam cupidō vir tulit ipse sinū
atque ita 'quid tenerōs lacrimīs corrumpis ocellōs?
     quod mātrī pater est, hoc tibi' dīxit 'erō.'
35 Rōmule, mīlitibus scīstī dare commoda sōlus:
     haec mihi sī dederis commoda, mīles erō.

3 *The poet displays his technique at the races in the Circus Maximus
(from the Amōrēs).*

He tries to chat up the girl sitting next to him.
'nōn ego nōbilium sedeō studiōsus equōrum;
     cui tamen ipsa favēs, vincat ut ille, precor.
ut loquerer tēcum, vēnī, tēcumque sedērem,
     nē tibi nōn nōtus, quem facis, esset amor.
5   tū cursūs spectās, ego tē: spectēmus uterque
     quod iuvat atque oculōs pāscat uterque suōs.
ō, cuicumque favēs, fēlīx agitātor equōrum!
     ergō illī cūrae contigit esse tuae?

19	**plausūs ... arte carēbant** their applause lacked discrimination = they applauded without discrimination, good and bad alike.
20	**praedae signa petīta** the eagerly-awaited (literally, 'sought for') sign (i.e. sign for seizing) their prey (i.e. the women).
21	**exsiliunt** they leap up. **animum ... fatentēs** declaring their love.
22	**īniciuntque**: translate the **-que** at the start of the line.
24	**agna novella** the little lamb. **vīsōs ... lupōs**: the lambkin runs off the moment it spots the wolf.
25	**sine lēge** without law and order = wildly.
26	i.e. they all changed colour.
27	i.e. they all felt the same panic but expressed it differently.
28	**pars ... pars ...** (one) part (of them) ... (another) part (of them) = some ... others .... **laniat** tears.
29	**altera ... altera** one ... another. **maesta** sad. **silet** is silent.
31	**repugnārat** (= **repugnāverat**): had fought back = fought back. **nimium** too much. **comitem negārat** (= **negāverat**) said no to her companion, i.e. the man who had seized her.
32	**sublātam cupidō ... sinū** lifted in his passionate embrace. **vir ... ipse** the man himself = the man without more ado.
33	**ita**: i.e. and as he did so. **corrumpis** do you spoil.
34	**quod mātrī pater est** what your father is to your mother.
35	**scīstī** (= **scīvistī**) **dare commoda** you knew the fringe benefits (**commoda**) to give (**scīstī dare** means literally 'you knew how to give'). **sōlus**: i.e. above all others, more than anyone else.
1	**studiōsus** (+ genitive) keen on. The poet says that it's not because he's keen on horses that he's here.
2	*precor tamen ut ille vincat cui ipsa favēs:* the subject of **favēs** is the girl the poet is sitting next to.
4	**quem facis ... amor** (my) love, which you are causing.
5	**cursūs** the races.
6	**pāscat**: **pāscō** (3) I feed.
7	**cuicumque** the antecedent is **agitātor equōrum** driver of horses, charioteer.
8	**ergō** and so. **illī ... contigit ...** has he the luck ...? **cūrae esse tuae** to be your care (predicative dative), i.e. to be cared for by you.

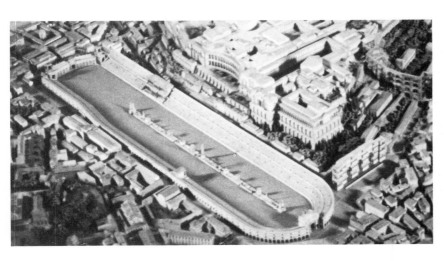

Circus Maximus

He dreams of what he would do if he were the charioteer backed by the girl.

hoc mihi contingat, sacrō dē carcere missīs
10 īnsistam fortī mente vehendus equīs
et modo lōra dabō, modo verbere terga notābō,
 nunc stringam mētās interiōre rotā;
sī mihi currentī fueris cōnspecta, morābor,
 dēque meīs manibus lōra remissa fluent.

The poet complains that the girl is trying to edge away from him, while other spectators are edging into her.

15 quid frūstrā refugis? cōgit nōs līnea iungī;
 haec in lēge locī commoda Circus habet.
tū tamen, ā dextrā quīcumque es, parce puellae:
 contāctū lateris laeditur illa tuī;
tū quoque, quī spectās post nōs, tua contrahe crūra,
20 sī pudor est, rigidō nec preme terga genū.

The race begins; the girl's charioteer proves a failure.

maxima iam vacuō praetor spectācula Circō
 quadriiugōs aequō carcere mīsit equōs.
cui studeās, videō; vincet, cuicumque favēbis:
 quid cupiās, ipsī scīre videntur equī.
25 mē miserum, mētam spatiōsō circuit orbe;
 quid facis? admōtō proximus axe subit.
quid facis, īnfēlīx? perdis bona vōta puellae;
 tende, precor, validā lōra sinistra manū.

The praetor starts the race

mētam circuit

9	**hoc mihi contingat** were this to be my luck. **sacrō de carcere missīs** released (literally, sent) from the sacred starting gate.
10	**īnsistam** (+ dative) I shall urge on. The poet is so carried away that he switches to the indicative, imagining that he is actually participating in the race. **vehendus** riding.
11	**modo . . . modo . . .** at one moment . . . at another moment . . . **lōra dabō** I shall give the reins = I shall give the horses their head.
12	**stringam mētās interiōre rotā** I shall graze the turning posts with the nearside wheel. At the ends of the stadium were the turning posts (**mētae**), three stone pillars with pointed tops. The charioteer's art lay in getting round these as closely as possible. Going around them too widely could waste valuable time. To bunch up with the other chariots would be to risk collision, and to strike rather than graze the posts would crash the chariot and lead to disaster.
13	**mihi** by me (dative of the agent). **fueris cōnspecta**: you shall have been caught sight of = if you are caught sight of.
14	**lōra** reins. **remissa** slack.
15	**refugis** do you back away? **līnea** the line, i.e. the groove in the stone which marked off the individual seats in the theatre.
16	**haec in lēge locī commoda** these advantages in the rules of the place.
17–20	**tū . . . tū . . .**: the poet addresses other spectators. **contāctū** (+ genitive) by contact with . . . . **post** behind. **tua contrahe crūra** pull your legs back. **rigidō . . . genū** with your bony knee.
21	**maxima . . . spectācula**: i.e. the races, the big event. **iam vacuō . . . Circō** the Circus (being) now empty. The procession which preceded the races is over. **praetor**: the praetor presided over the games. He began the races by dropping a white cloth.
22	**mīsit**: the verb has two objects: the praetor 'started' the **maxima spectācula** and 'let out' the teams of four horses (**quadriiugōs equōs**) from the starting gate, which ensures a fair start (**aequō**).
23	*vidēo cui studeās* **studeās** is in the subjunctive in indirect question; *(is) cuicumque favēbis vincet*.
25	**mē miserum**: (accusative of indignant exclamation) unhappy me = for God's sake, look at that! **mētam** the turning post. **spatiōsō** wide. **circuit** he is rounding.
26	**proximus** the man behind. **axe** with his axle. **subit** comes close.
28	**tende** pull tight. **validā** strong. **lōra** the reins.

The spectators call for a re-start; the girl's charioteer wins the
prize; will the poet win his prize?

       fāvimus ignāvō. sed enim revocāte, Quirītēs,
30   et date iactātīs undique signa togīs.
       ēn revocant; at, nē turbet toga mōta capillōs,
         in nostrōs abdās tē licet usque sinūs.
       iamque patent iterum reserātō carcere postēs,
         ēvolat admissīs discolor agmen equīs.
35   nunc saltem superā spatiōque īnsurge patentī:
         sint mea, sint dominae fac rata vōta meae.
       sunt dominae rata vōta meae, mea vōta supersunt;
         ille tenet palmam, palma petenda mea est.'
       rīsit et argūtīs quiddam prōmisit ocellīs:
40   'hoc satis hīc; aliō cētera redde locō.'

V

celeber, celebris, celebre	crowded
faciēs, faciēī, *f.*	appearance, face
frōns, frondis, *f.*	leaves, foliage
lupus, -ī, *m.*	wolf
nemus, nemoris, *n.*	grove
pateō, patēre, patuī	I am open
paveō, pavēre	I tremble, am afraid
pendeō, pendēre, pependī	I hang
plausus, -ūs, *m.*	applause
pudor, pudōris, *m.*	sense of shame, modesty
venia, veniae, *f.*	pardon
vēnor (1)	I hunt

G Word building

Make sure that you know the following compounds of
sequor, sequī, secūtus

assequor	I follow, attain
cōnsequor	I overtake, attain; I catch up
exsequor	I follow to the end, accomplish
īnsequor	I follow
persequor	I pursue to the end; I punish; I perform
subsequor	I follow after; I follow closely

29 **sed enim revocāte** but come on, call them back. **Quirītēs** Romans.

30 **iactātīs . . . togīs** by waving your togas. By doing this, the spectators give the sign (**signa**) that a re-start is necessary.

31 **ēn** look. **nē turbet toga mōta capillōs** so that the movement of (all) the togas doesn't disturb (your) hair.

32 **tē licet . . . abdās** you can hide. **in nostrōs usque sinūs** deep in the folds of my toga.

33–4 **patent . . . reserātō carcere postēs** the gates are open, with the starting-box unlocked. **admissīs** released. **discolor** of different colours. The charioteers wore the colours of their sporting party. These colours, red, green, white and blue, won fanatical support.

35 **saltem** at least. **spatiō** space. **īnsurge** (+ dative) rise into = make for (the open space).

36 *fac (ut) mea dominae(que) meae vōta rata sint.* **rata** fulfilled.

37 **supersunt** are left = are still to be fulfilled.

38 **ille** i.e. the winner. **palmam** the palm of victory.

39 **rīsit** she smiled. Until this line the poem has been a monologue; the poet has been talking the whole time. **argūtīs . . . ocellīs** with her lovely bright eyes (the diminutive **ocellīs** conveys the poet's emotion). **quiddam** something.

40 **hoc satis (est) hīc. cētera redde** pay the rest (of your promises). The girl's smile prompts the poet to hope for his reward after they have left the Circus.

G Numbers

Learn the following

singulī	one each	**semel**	once
bīnī	two each	**bis**	twice
ternī	three each	**ter**	three times
quaternī	four each	**quater**	four times
quīnī	five each	**quīnquiēns**	five times
sēnī	six each	**sexiēns**	six times
septēnī	seven each	**septiēns**	seven times
octōnī	eight each	**octiēns**	eight times
novēnī	nine each	**noviēns**	nine times
dēnī	ten each	**deciēns**	ten times

G Verbs followed by an infinitive

Some verbs are incomplete in meaning without an infinitive, e.g.
possumus **hoc facere,** *nōluērunt* **abīre,** *mālō* **lūdere.**
Revise the following

audeō, -ēre, ausus	I dare	**nōlō, nōlle, nōluī**	I do not wish,
coepī, -isse	I begin		am unwilling
cōgō, -ere, coēgī, coāctum	I force	**possum, posse, potuī**	I am able
cōnor (1)	I try	**sinō, -ere, sīvī, situm**	I allow
cōnstituō, -ere, -stituī, -stitūtum	I decide	**soleō, -ere, solitus**	I am accustomed
		videor, vidērī, vīsus	I seem
cupiō, cupere, cupīvī, cupītum	I desire	**volō, velle, voluī**	I wish, am willing
dēbeō (2)	I ought		
dēsinō, -ere, dēsiī	I cease		
discō, -ere, didicī	I learn		
doceō, -ēre, docuī, doctum	I teach		
incipiō, -cipere, -cēpī, -ceptum	I begin		
mālō, mālle, māluī	I prefer		

Ovid's longest work and, as many think, his finest, was the
Metamorphōsēs (= Changes of Shape). This is a great cycle of myths
and stories, linked together in an immensely long narrative of fifteen
books, in which the only common feature is that each story ends with a
change of shape. The following story comes at the end of the eighth
book:

Exercise 19.1

Read the following and then answer the questions after it

Baucis and Philemon entertain the gods, unawares.

Long, long ago Jupiter and Mercury came down to earth
disguised as mortals, to see how men behaved themselves. They
visited a thousand homes, seeking rest and food; a thousand homes
were barred against them. Finally they came to a poor cottage, where
a humble old couple lived, called Baucis and Philemon.

When the gods stooped to enter this lowly home, Philemon
invited them to sit and rest, while Baucis lit a fire with trembling hands
and put on a pot. She put in a cabbage which her husband had brought
from the garden; he took down a chine of smoked bacon and cut off a
piece from the pork they had kept so long. Baucis set the table with
fruit and nuts and honeycomb; Philemon brought in a jar of modest
wine. While the meal was cooking, they entertained their guests with
cheerful talk.

inter'ea toti'ens haustum cr'at'era repl'er'i
sponte su'a per s'eque vident succr'escere v'ina:
attonit'i novit'ate pavent manibusque sup'in'is
concipiunt Baucisque prec'es timidusque Phil'em'on
5 et veniam dapibus n'ull'isque par'atibus 'orant.
'unicus 'anser erat, minimae cust'odia v'illae,
quem d'is hospitibus domin'i mact'are par'abant.
ille celer penn'a tard'os aet'ate fat'igat
'el'uditque di'u, tandemque est v'isus ad ips'os
10 c'onf'ugisse de'os. super'i vetu'ere nec'ar'i
'd'i' que 'sumus, merit'asque luet v'ic'inia poen'as
impia,' d'ix'erunt; 'v'ob'is imm'unibus h'uius
esse mal'i dabitur. modo vestra relinquite t'ecta
ac nostr'os comit'ate grad'us, et in ardua montis
15 'ite simul!' p'arent amb'o, bacul'isque lev'at'i
n'ituntur long'o vest'igia p'onere cl'iv'o.

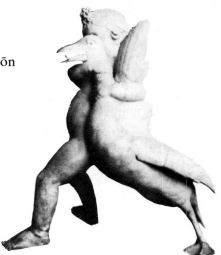

'unicus 'anser

1–2	**toti'ens haustum cr'at'era** the mixing-bowl (**cr'at'era** is acc. sing.), as often as it was drained. **repl'er'i** and **succr'escere** (was welling up) are infinitives dependent on **vident.** **per s'e** of itself.
3	**attonit'i novit'ate** astonished at the strangeness. **manibus sup'in'is** with palms turned upwards: the attitude of prayer.
4	**concipiunt** utter. **-que . . . -que . . .** both . . . and . . . . **timidus** describes both Baucis and Philemon.
5	**dapibus n'ull'isque par'atibus** for the meal and no preparations = for the fact that they had had to improvise the meal.
6	**'unicus 'anser** a single goose. **cust'odia** the guardian.
7	**d'is hospitibus** for the gods who were their guests.
8	**celer penn'a** swift with (its) wing = swift of wing. **tard'os aet'ate** Baucis and Philemon slow with age.
9	**'el'udit** eluded (them).
10	**c'onf'ugisse: confugi'o** I flee for refuge to. **super'i** the gods. **nec'ar'i** understand 'the goose' as the subject of this infinitive.
11–12	**merit'as luet v'ic'inia poen'as** your impious neighbours (literally, the impious neighbourhood) will pay the penalty they deserve.
12–13	**v'ob'is imm'unibus huius/esse mal'i dabitur**: literally, it will be granted to you to be immune from this evil (i.e. punishment). **modo** only. **t'ecta** house (poetic plural).
14	**comit'ate** accompany. **ardua** the heights.
15	**simul**: i.e. **cum n'ob'is.** **bacul'is lev'at'i** propped up on their sticks.
16	**n'ituntur** they struggle. **cl'iv'o** on the slope.

When they were an arrow's flight from the top, they looked back and saw the whole countryside submerged in water; only their own house was left. While they watched in amazement and dismay, their little cottage was turned into a temple. Marble columns took the place of the wooden roof props; the straw thatch grew yellow and became a roof of gold.

Then Mercury said, 'Tell us, good old man and wife, worthy of your good husband, what do you wish?' After Philemon had consulted Baucis a moment, he said, 'We ask to become your priests and the guardians of your temple, and, since we have lived our lives together in harmony, may the same hour carry us off together.' Their prayer was granted. They were the guardians of the temple while life was allowed them.

When they were very old and were standing before the temple steps telling over the story of the place, Baucis saw Philemon sprouting leaves and old Philemon saw Baucis sprouting too. As the tree tops grew over their two faces, while they still could, they spoke to each other; each said,

'Farewell, dear wife/husband,*' just as the bark covered and hid their mouths.

To this day, a peasant of the place will point to two trees close to each other, growing from a single trunk.

*coniūnx: the Latin word means either 'wife' or 'husband' (literally 'mate').

1 What is the first thing that makes Baucis and Philemon realize that their guests have supernatural powers?
2 How do they feel when they realize this?
3 What do they wish to do to show that they know they have divine guests? Why is this such an important gesture for them?
4 Say what the intended victim does.
5 Why did the gods tell Baucis and Philemon to go to the top of the hill?
6 What evidence is there in the Latin narrative that Baucis and Philemon are old?
   For the last two questions, you may use the parts of the tale given in English
7 What is the message – or moral – of the story as a whole?
8 This is a tale of old world hospitality. How serious is its tone?

Exercise 19.2

Translate into Latin

1 You do want to go to the Circus Maximus, don't you?
2 Let's sit (down). We shall be able to see the chariots well from here.
3 If I had stayed at home, I should not have seen the famous charioteer
   (**agitātor**).
4 After seeing that race, I returned home happily.
5 When I arrived at my house, I found Ovidius who was waiting by the door.

Exercise 19.3

Translate into Latin

Yesterday my husband and I decided to go to the Circus Maximus to
see the contests. When we arrived there, so many people were present
that we could scarcely find a place. At last we sat next to a man who
began to talk to me about the charioteers (**agitātor**). 'Who do you
think will win?' he asked; and he said many other things. I smiled at
him, but I did not want to speak to him because he seemed to be very
impudent (**impudēns**). When I had returned home with my husband, I
decided not to go to the Circus again. I have been there once this year.
I shall not go twice.

CHAPTER XX

Ovid in exile

*The reasons for Ovid's banishment remain mysterious. The poet himself refers to a poem (**carmen**) and a mistake (**error**). We are unlikely ever to find out exactly what the mistake was. He may have innocently overheard part of a conspiracy against the Emperor and done nothing about it.*

On the subject of the poem, however, we are on surer ground. Augustus was eager to improve the lax moral standards of the Rome of his day. Yet in AD 2 he had to banish his only daughter Julia for flagrant adultery and six years later Julia's daughter had to be banished too, also for immorality. The Emperor would have felt little affection for the poet who had so often encouraged the sexual permissiveness which he himself had been at such pains to suppress. All of Ovid's love poetry must have seemed a provocation to him, but the <u>Ars Amātōria</u> – from which you have read an excerpt – would have struck him as especially offensive. When he banished Ovid, Augustus had this poem banned from Rome's three public libraries.

Ovid left for exile in Tomis on the Black Sea in December AD 8 and had arrived there by the following autumn. He lived in this frontier town inhabited by half-bred Greeks and barbarian Getae until his death in AD 17.

He had a thoroughly uncomfortable voyage out, only just weathering the winter storms.

      dī maris et caelī – quid enim nisi vōta supersunt? –
          solvere quassātae parcite membra ratis.
      mē miserum, quantī montēs volvuntur aquārum!
          iam iam tāctūrōs sīdera summa putēs.
5   quantae dīductō subsīdunt aequore vallēs!
          iam iam tāctūrās Tartara nigra putēs.
      quōcumque aspiciō, nihil est, nisi pontus et āēr,
          flūctibus hic tumidus, nūbibus ille mināx.
      inter utrumque fremunt immānī murmure ventī.
10   nescit, cui dominō pāreat, unda maris.
      rēctor in incertō est nec quid fugiatve petatve
          invenit: ambiguīs ars stupet ipsa malīs.
      scīlicet occidimus, nec spēs est ūlla salūtis,
          dumque loquor, vultūs obruit unda meōs.

Statue of Ovid in a square at Tomis

1   **nisi** except for.   **supersunt** are left. In English we should say 'what is left . . . ?'.

2   **solvere . . . parcite** spare to loosen = I beg you not to break.   **quassātae** shattered, agreeing with **ratis**.   **membra** the limbs = the frame.

3   **mē miserum** unhappy me = God help me! (accusative of exclamation).

4   **tāctūrōs (esse)**.   **putēs** you would think.

5   **dīductō** torn apart.   **subsīdunt** sink beneath (us).

6   **Tartara** Tartarus, i.e. the Underworld.

7   **pontus** the sea.

8   **tumidus** swollen.   **mināx** threatening.

9   **fremunt** roar.

10   **cui dominō**: i.e. which of the various winds.

11   **rēctor** the helmsman.   **quid fugiat . . .** what to avoid, steer clear of.

12   **ambiguīs** baffling.   **ars ipsa** his skill itself = even his great skill.

13   **scīlicet** surely.

14   **vultūs . . . meōs** my face.   **obruit** overwhelms.

When he got there, he found himself living under conditions of extreme discomfort; he writes graphically of the horrors of the climate at Tomis:

It is terrible here even in the summer.

15 sīquis adhūc istīc meminit Nāsōnis adēmptī,
    et superest sine mē nōmen in urbe meum,
suppositum stellīs numquam tangentibus aequor
    mē sciat in mediā vīvere barbariā.
Sauromatae cingunt, fera gēns, Bessīque Getaeque,
20   quam nōn ingeniō nōmina digna meō!
dum tamen aura tepet, mediō dēfendimur Histrō:
    ille suīs liquidus bella repellit aquīs.

mediō dēfendimur Histrō

The rigours of winter.

at cum trīstis hiems squālentia prōtulit ōra,
    terraque marmoreō est candida facta gelū,
25 nix iacet, et iactam nec sōl pluviaeve resolvunt,
    indūrat Boreās perpetuamque facit.
tantaque commōtī vīs est Aquilōnis, ut altās
    aequet humō turrēs tēctaque rapta ferat.
pellibus et sūtīs arcent mala frīgora brācīs,
30   ōraque dē tōtō corpore sōla patent.
saepe sonant mōtī glaciē pendente capillī,
    et nitet inductō candida barba gelū.

All of Nature is gripped by the ice.

quid loquar ut vinctī concrēscant frīgore rīvī,
    dēque lacū fragilēs effodiantur aquae?
35 quāque ratēs ierant, pedibus nunc ītur, et undās
    frīgore concrētās ungula pulsat equī;
vīdimus ingentem glaciē cōnsistere pontum,
    lūbricaque immōtās testa premēbat aquās.
inclūsaeque gelū stābunt in marmore puppēs,
40   nec poterit rigidās findere rēmus aquās.
vīdimus in glaciē piscēs haerēre ligātōs,
    sed pars ex illīs tum quoque vīva fuit.

218

15	**adhūc** still. **istīc** there, i.e. in Rome. **Nāsōnis adēmptī** the Naso (i.e. the Ovid) Rome has lost. Ovid's full name was Publius Ovidius Naso.
16	**superest** is left, survives.
17–18	**suppositum stellīs**: placed (dwelling) beneath the stars (**suppositum** agrees with **mē**). Ovid suggests the cold remoteness of Tomis by referring to the northern constellations such as the Great and Lesser Bear which never sink below the sea. **barbariā** (ablative) the barbarian world.
19	These are the names of three local tribes.
20	*nōmina quam nōn digna*.
21	**tepet** is warm. **mediō dēfendimur Histrō** we are defended (from the hostile tribes) by the river Hister in the middle, i.e. acting as a barrier. The Hister is the Danube.
22	**ille . . . liquidus** he (i.e. the river) as he flows.
23	**squālentia prōtulit ōra** has thrust his rough face forth. Like the river, the winter is personified.
24	**marmoreō . . . gelū** with frost as white and hard as marble.
25	**iactam (nivem)**: the thrown down snow, i.e. the snow once it has fallen. **nec sōl pluviaeque resolvunt** neither sun nor rain can melt . . . .
26	**indūrat Boreās** the North wind hardens (it). **perpetuam** everlasting.
27	**commōtī . . . Aquilōnis** of the North wind (when) set in motion = of the North wind when it is blowing violently.
28	**aequet** levels = flattens. **tēcta** houses. **ferat** carries away.
29	**arcent** they keep out. **pellibus . . . et sūtīs . . . brācīs** with skins and sewn trousers = with trousers of sewn skins (hendiadys).
30	**patent** lie open, are exposed.
31	**sonant** make(s) a noise, tinkle(s). **capillī** (their) hair.
32	**inductō . . . gelū** frost having been drawn over it = with a layer of frost.
33	**quid loquar ut . . .** why should I tell how . . . . *rīvī, frīgore vinctī, concrēscant* the streams, congealed by the cold, are frozen over.
34	**fragilēs** brittle. It is not in fact water but ice which has to be broken. **effodiantur** are dug out of.
35	**quā** where. **ītur** it is gone = men go.
36	**concrētās** congealed. **ungula** the hoof. **pulsat** strikes.
37	**pontum** sea.
38	**lūbrica . . . testa** a slippery shell = a coating of ice. **premēbat** was holding.
39	**gelū** by the cold. **stābunt**: Ovid uses the future tense – he is describing something that will (i.e. is liable to) happen. **in marmore** on the marble (surface of the sea).
40	**rigidās** stiff. **findere** to cut.
41	**haerēre ligātōs** stick fast, embedded in . . . .
42	**pars** some (literally: part) of the fish.

The enemy attacks over the ice.
prōtinus aequātō siccīs Aquilōnibus Histrō
    invehitur celerī barbarus hostis equō;

45   hostis equō pollēns longēque volante sagittā
    vīcīnam lātē dēpopulātur humum.
diffugiunt aliī; nūllīsque tuentibus agrōs
    incustōdītae dīripiuntur opēs.
pars agitur vinctīs post tergum capta lacertīs,
50   respiciēns frūstrā rūra Laremque suum:
pars cadit hāmātīs miserē cōnfixa sagittīs:
    nam volucrī ferrō tinctile vīrus inest.
quae nequeunt sēcum ferre aut abdūcere, perdunt,
    et cremat īnsontēs hostica flamma casās.

hostis equō pollēns

Paralysed by fear, the natives neglect the land.
All lies desolate and barren.
55   tunc quoque cum pāx est, trepidant formīdine bellī,
    nec quisquam pressō vōmere sulcat humum.
aut videt aut metuit locus hic, quem nōn videt, hostem;
    cessat iners rigidō terra relicta sitū.
aspicerēs nūdōs sine fronde, sine arbore, campōs:
60   heu loca fēlīcī nōn adeunda virō!
ergō tam lātē pateat cum maximus orbis,
    haec est in poenam terra reperta meam.

Inevitably under such conditions, Ovid fell ill and wrote a sad
letter to his wife:
haec mihi sī cāsū mīrāris epistula quārē
    alterius digitīs scrīpta sit, aeger eram.
65   aeger in extrēmīs ignōtī partibus orbis,
    incertusque meae paene salūtis eram.
nec caelum patior, nec aquīs adsuēvimus istīs,
    terraque nescio quō nōn placet ipsa modō.
non domus apta satis, nōn hīc cibus ūtilis aegrō,
70   nūllus, Apollineā quī levet arte malum,
nōn quī sōlētur, nōn quī lābentia tardē
    tempora nārrandō fallat, amīcus adest.
lassus in extrēmīs iaceō populīsque locīsque,
    et subit adfectō nunc mihi, quicquid abest.
75   omnia cum subeant, vincis tamen omnia, coniūnx,
    et plūs in nostrō pectore parte tenēs.
tē loquor absentem, te vōx mea nōminat ūnam;
    nūlla venit sine tē nox mihi, nūlla diēs.

43	**aequātō** levelled. **siccīs** dry = freezing.
44	**invehitur** rides to the attack.
45	**pollēns** powerful. The enemy is noted for his prowess both on horseback and in archery.
46	**vīcīnam** neighbouring. **dēpopulātur** lays waste.
48	**incustōdītae** unguarded. **opēs** *f. pl.* resources.
49	**pars agitur . . . capta** some are driven off into captivity. **vinctīs . . . lacertīs** their arms bound.
50	**Larem** their household god = their homes.
51	**hāmātīs** barbed. **cōnfīxa** transfixed.
52	**volucrī** (ablative) winged. The 'winged iron' refers to arrows. **tinctile vīrus** poison smeared (on them).
53	**nequeunt** they cannot.
54	**cremat** burns. **īnsontēs . . . casās** their unoffending huts. **hostica** the enemy's.
55	**trepidant** they are disturbed. **formīdine** with dread of.
56	**sulcat** ploughs. **pressō vōmere** with down-pressed ploughshare.
57	**quem**: antecedent **hostem.**
58	**cessat iners** lies idle. **rigidō sitū** in unbroken neglect = in neglect, unbroken by the plough.
59	**aspicerēs** you may see = one may see.
61	*ergō, cum maximus orbis (= the world) tam lātē pateat* (**cum** although).
62	**est . . . reperta** has been discovered. **in poenam . . . meam** for my punishment.
63–4	*sī cāsū mīrāris quārē haec mea epistula digitīs alterius scrīpta sit.* **digitīs** with the fingers. **aeger eram** I was ill (when this letter was written) = I am ill.
65	**ignōtī orbis** of an unknown world.
67	**caelum** the climate. **adsuēvimus** have we (plural for singular) become accustomed.
68	**nescio quō . . . modō** I don't know how = in some way I cannot explain.
69	**apta (aegrō).**
70	**nūllus . . . quī levet** no-one who may relieve = no-one to relieve. **Apollineā . . . arte**: Apollo's art was medicine.
71–2	**quī sōlētur** who may comfort (me) = to comfort me. **quī**: antecedent **amīcus** – there is no friend here to make the time pass more quickly.
73	**lassus** weary, faint.
74	*quicquid abest nunc subit mihi* (comes to my mind) *adfectō* (affected (with sickness), i.e. in my sickness).
75	*cum* (although) *omnia subeant* (come to (my) mind).
76	**plūs . . . parte** more than half (a share) (**pars** (here) = share).
77	**nōminat** calls by name.

He kept himself going by writing poetry but complains that he has almost forgotten how to speak Latin.

dētineō studiīs animum fallōque dolōrēs,

80   experior cūrīs et dare verba meīs.

quid potius faciam dēsertīs sōlus in ōrīs,

quamve malīs aliam quaerere cōner opem?

sīve locum spectō, locus est inamābilis, et quō

esse nihil tōtō trīstius orbe potest;

85   sīve hominēs, vix sunt hominēs hōc nōmine dignī,

quamque lupī saevae plūs feritātis habent.

in paucīs remanent Grāiae vestīgia linguae,

haec quoque iam Geticō barbara facta sonō.

ūnus in hōc nēmō est populō, quī forte Latīnē

90   quaelibet ē mediō reddere verba queat.

ille ego Rōmānus vātēs – ignōscite, Mūsae! –

Sarmaticō cōgor plūrima mōre loquī.

Ⅴ

ambō, ambae, ambō	both
bōs, bovis, *m. & f.*	ox, cow
ecce!	look, behold
ferus-a-um	wild, savage
frōns, frontis, *f.*	forehead
horreō, horrēre, horruī	I bristle, shudder (at); I stand in awe of
humus, -ī, *f.*	ground
piscis, piscis, *m.*	fish
puppis, puppis, *f.*	stern, poop; ship
ratis, ratis, *f.*	raft; ship
requiēscō, -ere, requiēvī	I rest, relax
sagitta, -ae, *f.*	arrow
tueor, tuērī, tuitus	I protect, look at
vinciō, vincīre, vīnxī, vīnctum	I bind

Ⅾ A number of verbs of the third and mixed conjugations also have first conjugation forms. Learn the following. The meaning is often slightly different and we have given it in such cases

canō	**cantō** (1)	**vertō**	**versō** (1)
capiō	**captō** (1) I strive after	**speciō** (I look at)	**spectō** (1) I look at,
currō	**cursō** (1) I run hither and thither		contemplate
dīcō	**dictō** (1) I dictate		
iaciō	**iactō** (1)		
ostendō	**ostentō** (1)		
sequor	**sector** (1) I follow eagerly, attend		

222

79	**dētineō** i.e. 'I keep busy'.
80	*et experior* (I try) *dare.* **cūrīs ... dare verba meīs** to cheat my cares.
81	**quid potius** what rather? = what else. **dēsertīs** desolate.
83	**inamābilis** unlovely. **quō** (a place) than which.
85	**sīve hominēs (spectō).** **hōc nōmine**: i.e. the name of man.
86	**saevae plūs feritātis** more (of) savage fierceness.
87	**remanent** there remain, still exist. **Grāeiae ... linguae** of the Greek language.
88	**facta (sunt).**
89–90	**quī ... queat** who by chance could utter in Latin any colloquial language at all. **forte** (= by chance) emphasizes the remoteness of the possibility. **ē mediō verba** colloquial language, i.e. the natives speak the language, if at all, in stilted pigeon Latin. **quaelibet ... verba**: any words at all.
91	**ille vātēs** the famous bard.

❻ Words followed by the genitive

accūsō (1)	I accuse someone (acc.) of something (gen.)
admoneō (2)	I remind someone (acc.) of something (gen.)
meminī, meminisse*	I remember
misereor (2)	I pity
oblīvīscor, oblīvīscī, oblītus*	I forget

*can be followed by accusative

avidus-a-um	eager for
cupidus-a-um	desirous of
incertus-a-um	uncertain of
īnscius-a-um	ignorant of
memor, memoris	mindful of
immemor, immemoris	forgetful of
perītus-a-um	skilled in

Exercise 20.1

Translate

1 Rōmae semper memor, Ovidius etiam nunc eō redīre vult.

2 Aenēās negāvit sē unquam Dīdōnis oblītūrum esse.

3 Ovidius, quamquam artis dīcendī perītissimus erat, Augustō persuādēre nōn potuit ut suī miserērētur.

4 Nāsōnis meminī, lūsōris illius tenerōrum amōrum.

5 Dīdō Aenēān perfidiae accūsāvit, quia ille immemorem fideī sē praestābat.

Exercise 20.2

Read the following passage carefully and then answer the questions after it.

When Ovid was twenty-four, his fellow poet Albius Tibullus died.
Ovid wrote this poem in tribute to him.

Memnona sī māter, māter plōrāvit Achillem,
  et tangunt magnās trīstia fāta deās,
flēbilis indignōs, Elegēia, solve capillōs:
  ā nimis ex vērō nunc tibi nomen erit.
5 ille tuī vātēs operis, tua fāma, Tibullus,
  ārdet in exstrūctō corpus ināne rogō.
cum rapiunt mala fāta bonōs, (ignōscite fassō)
  sollicitor nūllōs esse putāre deōs.
vīve pius: moriēre pius; cole sacra: colentem
10 mors gravis ā templīs in cava busta trahet.
carminibus cōnfīde bonīs: iacet ecce Tibullus;
  vix manet ē tōtō, parva quod urna capit.
sī tamen ē nōbīs aliquid nisi nōmen et umbra
  restat, in Elysiā valle Tibullus erit.
15 obvius huic veniēs hederā iuvenālia cinctus
  tempora cum Calvō, docte Catulle, tuō;
hīs comes umbra tua est, sī qua est modo corporis umbra;
  auxistī numerōs, culte Tibulle, piōs.
ossa quiēta, precor, tūtā requiēscite in urnā,
20 et sit humus cinerī nōn onerōsa tuō.

parva quod urna capit

A Roman funeral

1ff.	i.e. if goddesses can mourn the death of their children, then you, Elegy, should mourn Tibullus. Elegeia, the spirit of elegiac poetry, is personified as a goddess whose children are the elegiac poets.
1	**Memnona**: accusative case. The Ethiopian prince Memnon, son of Aurora (the dawn), was killed at Troy by Achilles. His mother wept tears of dew for him every morning. Achilles, son of Thetis, a sea goddess, was killed later by Paris. **plōrāvit** wept for.
3	**flēbilis** tearful (agreeing with **Elegēia**). **indignōs capillōs** your hair which has not deserved such treatment. Women loosed their hair in mourning.
4	**nimis ex vērō** only too true. Ovid refers to the derivation of the name Elegy from the Greek **e legei** ((he) cries woe).
5	**vātēs** bard. **tua fama**: i.e. the man who made you (Elegy) famous.
6	**in exstrūctō . . . rogō** on a high (literally, built up) pyre.
7	**fassō**: (dative) my confession.
8	**sollicitor** I am moved to.
9	**moriēre = moriēris.**
10	**in cava busta** into a hollow tomb.
11	**cōnfīde**: the tone is ironical.
12	**parva urna**: after cremation, the bones and ashes of the dead were stored in an urn.
13	**umbra** shadow, shade.
14	**restat** remains. **Elysiā** of Elysium. Elysium was the dwelling-place of the good in the Underworld.
15–16	**obvius** (+ dative) to meet. **hederā iuvenālia cinctus/tempora** bound as to your youthful temples with ivy = your youthful temples bound with ivy. Calvus was a friend and fellow poet of Catullus (see Chapters 9–11). **docte**: this means more than 'learned' – perhaps translate 'the great craftsman'.
17	**sī . . . modo** if only.
18	**numerōs** verse, poetry, **culte** cultivated, civilized.
19	**ossa** his bones.
20	**onerōsa** burdensome to = a heavy weight on.

1 Why is Elegy asked to mourn?
2 Why does Ovid feel that the gods may not exist (1.8)?
3 Translate ll.9–12.
4 Do you have the impression from this poem that Ovid believes in a full afterlife in the Elysian fields?
5 How serious is this poem? How deep do you feel Ovid's grief to be? How moving do you find this tribute?
6 Compare this poem with Catullus's elegy on his brother (p.126). Which poem do you prefer, and why?

Exercise 20.3

Translate into Latin

1 Ovid's wife asked the emperor to forgive him.
2 Poets should write poetry about wars, not about love.
3 If ever I return to Rome, I shall lead a very respectable life.
4 He asked where Corinna's parrot had come from.
5 When the parrot had died, Corinna was very sad.

Exercise 20.4

Translate into Latin

Pyramus and Thisbe, the one the most handsome of young men, the other more beautiful than all the girls of the East (**Oriēns**), were accustomed to speak through a chink (**rīma**) in the wall. After deciding to leave their homes at night, they said that they would meet under the shade of a mulberry tree (**mōrus, -ī,** *f.*). When a lioness (**leaena**) came to drink water from the fountain, Thisbe was wounded and fled into a cave. Pyramus, however, seeing her clothing on the ground stained (**tinctus-a-um**) with blood, believed that she was dead. 'If I had come before,' he said, 'I would have saved her.' He was so unhappy that he killed himself. When Thisbe returned to the tree, she found the body of her lover and she too killed herself. The tree's berries (**bāca, -ae,** *f.*), which before had been white, ever since that time have been purple.

Brief summary of Latin syntax

1 Main clauses

(a) The *indicative* is used in statements and questions:
Cicerō hoc fēcit Cicero did this.
quid fēcit Cicerō? What did Cicero do?

(b) the *imperative* is used in commands:
hoc fac Do this. **nōlī hoc facere** Don't do this.

(c) The *subjunctive* is used:

(i) in exhortations and commands (jussive), negative **nē:**
nē hoc faciāmus Let us not do this.
hoc faciās Do this.
nē hoc fēcerīs Don't do this.

(ii) in wishes (negative **nē** or **nōn**):
nē hoc faciās/utinam nē hoc faciās May you not do this!
utinam nē hoc facerēs I wish you were not doing this.
utinam hoc nōn fēcissēs I wish you had not done this.

(iii) in deliberative questions:
quid faciāmus? What are we to do?

(iv) in potential clauses:
hoc facere velim/nōlim/mālim I should like/not like/ prefer to do this.

(v) in impossible and remote (future) conditional clauses:
sī hoc fēcissēs, stultus fuissēs If you had done this, you would have been foolish.
sī pater adesset, nōs iuvāret If father were here, he would be helping us.
sī hoc faciās, stultus sīs If you were to do this, you would be foolish.

2 Subordinate clauses

(a) The *indicative* is used in:

(i) definite relative clauses:
haec est casa quam aedificāvit Iacobus This is the house which Jack built.

(ii) definite temporal clauses:
ubi/postquam/ut Iacobus casam aedificāvit, Iūliam dūxit When Jack had built the house, he married Julia.

(iii) causal clauses:
quod/quia/quoniam Iacobus casam nōn cōnfēcerat, Iūlia īrāta est Because Jack had not finished the house, Julia was angry.
(on **quod,** see also (b) ix)

(iv) concessive clauses with **quamquam:**
quamquam Iacobus casam bene aedificāverat, ventus eam obruit Although Jack had built the house well, the wind knocked it down.

(v) open conditional clauses:

sī hoc fēcistī, stultus fuistī
If you did this you were foolish
sī hoc facis, stultus es
If you are doing this, you are foolish.
sī hoc fēceris, stultus eris
If you do this, you will be foolish.

(vi) **cum** (= when) takes the indicative in present and future time:
Iacobus, cum casam cōnfēcerit, Iuliam dūcet
When Jack finishes the house, he will marry Julia.
In past time **cum** (= when) usually takes the subjunctive, but it is found with the indicative under special circumstances (see Chapter 18, page 198).

(b) The *subjunctive* is used in:

(i) purpose clauses following **ut/nē**:
puer domum redit ut patrem videat
The boy is returning home to see his father.
puer domum rediit nē ā patre reprehenderētur
The boy returned home so that he would not be blamed by his father.

(ii) in indirect commands and requests:
puer patrem rogat ut sē iuvet The boy asks his father to help him.
pater fīliō imperāvit nē domō discēderet The father told his son not to leave home.

(iii) in indirect questions:

rogat	**faciam**	is asking	I am doing.
amīcus rogābit quid fēcerim		My friend will ask	what I did.
rogāvit	**factūrus sim**	has asked	I am going to do.
rogābat	**facerem**	was asking	I was doing.
amīcus rogāvit quid fēcissem		My friend asked	what I had done.
rogāverat	**factūrus essem**	had asked	I was going to do.

(This table illustrates the full scheme of the sequence of tenses, which applies to all subjunctives in subordinate clauses except for consequence clauses.)

(iv) after verbs of fearing with **nē** (negative **nē nōn**)
puer veritus est nē pater sibi īrāscerētur
The boy was afraid his father would be angry with him.

(v) Clauses of consequence, introduced by **ut** (negative **ut nōn**):
puer adeō timēbat ut aufūgerit
The boy was so afraid that he ran away.

(vi) **cum** (= when) usually takes the subjunctive in past time:
puer, cum domum rediisset, patrem quaesīvit
When the boy had returned home, he looked for his father.
cum (= since/although) always takes the subjunctive:
cum fīlius mātrem adiuvet, pater eum laudat
Since the son is helping his mother, father praises him.
cum puer fessus esset, dormīre tamen nōn potuit
Although the boy was tired, he could not sleep.

(vii) in relative clauses expressing purpose and consequence:

pater fīlium mīsit quī mātrem quaereret

The father sent his son to look for his mother.

servus dignus est quī praemium accipiat

The slave is worthy (= deserves) to receive a reward.

and in generic clauses:

nōn eī sunt quī pauperēs adiuvent

They are not the sort of people to help the poor.

(viii) in temporal clauses which express purpose as well as time:

exspectābam dum pater advenīret.

I was waiting for father to arrive.

aufūgimus antequam hostēs nōs caperent

We fled before the enemy could catch us.

(ix) in causal clauses with **quod** which give an alleged reason:

iuvenem condemnāvērunt quod argentum fūrātus esset

They condemned the young man because (as they said) he had stolen the money (= for stealing the money).

(x) in concessive clauses with **quamvīs** or **licet.**

quamvīs fessus sīs, tibi properandum est

However tired you may be, you must hurry.

licet fessus sīs, tibi festīnandum sit

Although you may be tired, you must hurry.

(xi) ALL subordinate clauses in indirect speech have their verbs in the subjunctive:

Direct speech: **'ego plūrimum Caesarī dēbeō, quod eius operā stīpendiō līberātus sum, quod Aduaticīs fīnitimīs meīs pendere cōnsuēveram, quodque fīlius ab Caesare remissus est, quem Aduaticī in servitūte tenēbant.**

'I owe very much to Caesar, because through his efforts I was freed from the tax which I used to pay to the Aduatici, my neighbours, and because my son was sent back by Caesar, whom the Aduatici were holding in slavery.'

Indirect speech: **dīxit sē plūrimum Caesarī dēbēre, quod eius operā stīpendiō līberātus esset, quod Aduaticīs fīnitimīs suīs pendere cōnsuēvisset, quodque fīlius ab Caesare remissus esset, quem Aduaticī in servitūte tenuissent.**

LATIN-ENGLISH VOCABULARY

This list contains all but the very commonest words. Hyphens are used for the purpose of abbreviation. They have no etymological significance.

ā/ab + abl. from, by
ab-dō, -ere, -didī, -ditum I put away; I hide
ab-rumpō, -ere, -rūpī, -ruptum I break off
absēns, absentis absent
ab-solvō, -ere, -solvī, -solūtum I acquit
absum, abesse, āfuī I am away from, absent
ac and
ac-cēdō, -ere, -cessī, -cessum I approach
ac-cendō, -ere, -cendī, -cēnsum I set on fire; I excite
accidit, -ere, accidit it happens
ac-cipiō, -ere, -cēpī, -ceptum I receive; I hear
acclāmāti-ō, -ōnis, f. shout, cat-call
accūsō (1) I accuse
āc-er, -ris, -re keen, fierce
aci-ēs, -ēī, f. line of battle
ac-quiēscō, -ere, -quiēvī I rest
ad + acc. to
ad-dō, -dere, -didī, -ditum I add
ad-dūcō, -ere, -dūxī, -ductum I lead to, bring
ad-eō, -īre, -iī, -itum I approach
adeō (adverb) so, to such an extent
adflictō (1) I damage
ad-flīgō, -ere, -flīxī, -flīctum I strike, dash
adhūc still
ad-iciō, -icere, -iēcī, -iectum I add
ad-igō, -ere, -ēgī, -āctum I drive to, bring to
ad/imō, -ere, -ēmī, -ēmptum I take away
adipīscor, -ī, adeptus I obtain, win
adit-us, -ūs, m. approach
ad-iuvō, -āre, -iūvī, -iūtum I help
administrō (1) I manage, govern
admīror (1) I admire, wonder at
ad-mittō, -ere, -mīsī, -missum I let in; I commit
admoneō (2) I warn
ad-olēscō, -ere, -olēvī I grow up
ad-orior, -īrī, -ortus I attack
adscēns-us, -ūs, m. ascent
ad-sequor, -ī, -secūtus I follow; I catch
ad-sum, -esse, -fuī I am present
adulēscēn-s, -tis, m. young man
adulēscenti-a, -ae, f. youth
adulter, -ī, m. adulterer
ad-veniō, -īre, -vēnī, -ventum I arrive
advent-us, -ūs, m. arrival
adversāri-us, -a, -um opposed to; an opponent
adversor (1) I oppose
advers-us, -a, -um opposite
advolō (1) I fly to
aedifici-um, -ī, n. building
aedificō (1) I build
aeg-er, -ra, -rum sick, ill

aegrē with difficulty, unwillingly; scarcely
aequor, -is, n. sea
aequ-us, -a, -um equal, level, fair
  aequē ac as much as
  aequō animō calmly
aes, aeris, n. bronze, money
aest-ās, -ātis, f. summer
aestimō (1) I value
aest-us, -ūs, m. tide
aet-ās, -ātis, f. age
aetern-us, -a, -um eternal
af-ferō, -ferre, attulī, adlātum I bring to; I report
af-ficiō, -ere, -fēcī, -fectum I affect
ager, agrī, m. field
agg-er, -eris, m. mound, rampart
ag-gredior, -gredī, -gressus I attack
ag-men, -minis, n. column
agō, -ere, ēgī, āctum I drive; I do
  agō dē I discuss
agricol-a, -ae, m. farmer
āit he said
alb-us, -a, -um white
aliquamdiū for some time
aliquis, aliquid someone, something
aliquot (indecl.) several
aliter otherwise
ali-us, -a, -ud other
  alius ac different from
alt-er, -era, -erum one or the other
altitūd-ō, -inis, f. height, depth
alt-us, -a, -um high, deep
  alt-um, -ī, n. the deep (sea)
amābil-is, -e lovable, amiable
amān-s, -tis loving, lover
amātor, -is, m. lover
amb-ō, -ae, -ō both
ambulō (1) I walk
amīciti-a, -ae, f. friendship
amīc-us, -ī, m. friend
ā-mittō, -ere, -mīsī, -missum I let slip, lose
amō (1) I love
amor, -is, m. love
ampl-us, -a, -um large
  amplius (adverb) more. longer
an? or
animal, -is, n. animal
ancor-a, -ae, f. anchor
angusti-ae, -ārum, f.pl. narrows, pass
angust-us, -a, -um narrow
anim-a, -ae, f. soul, life
animad-vertō, -ere, -vertī, -versum I notice, perceive

anim-us, -ī, *m.* mind, spirit
ann-us, -ī, *m.* year
ante + acc. before
anteā (adverb) before
antequam (conjunction) before
antīqu-us, -a, -um old
aper-iō, -īre, -uī, -tum I open
apert-us, -a, -um open
appāreō (2) I appear; I am clear
appellō (1) I call
ap-pellō, -ere, -pulī, -pulsum I drive to
appet-ō, -ere, -iī, -ītum I attack
apt-us, -a, -um suitable, fit
apud + acc. at, with
aqu-a, -ae, *f.* water
aquil-a, -ae, *f.* eagle, legionary standard
Aquil-ō, -ōnis, *m.* North wind
ār-a, -ae, *f.* altar
arbitror (1) I think
arbor, -is, *f.* tree
arceō (2) I keep at a distance
arcess-ō, -ēre, -īvī, -ītum I summon
ārdeō, -ere, ārsī I am on fire
ārdēn-s, -tis burning, passionate
ārdor, -is, *m.* passion
argent-um, -ī, *n.* silver, money
ārid-us, -a, -um dry
   ārid-um, -ī, *n.* dry land
arm-a, -ōrum, *n.pl* arms
armō (1) I arm
ars, artis, *f.* art, skill
art-us, -ūs, *m.* limb
arx, arcis, *f.* citadel
a-scendō, -ere, -scendī, -scēnsum I climb
ascēnsus = adscēnsus
asper, -a, -um rough, harsh, dangerous
aspiciō, -ere, aspexī, aspectum I look at
as-sequor = ad-sequor
astō, -āre, astitī I stand near
at but
atque and
ātri-um, -ī, *n.* hall
at-tingō, -ere, -tigī, -tāctum I touch
attollō, -ere I raise, lift up
auctor, -is, *m.* author, adviser
auctōrit-ās, -ātis, *f.* authority, influence
audāci-a, -ae, *f.* boldness, rashness
aud-āx, -ācis bold, rash
audeō, audēre, ausus I dare
auferō, auferre, abstulī, ablātum I carry away
augeō, -ēre, auxī, auctum I increase, enlarge
audiō (4) I hear, listen to
aur-a, -ae, *f.* air, breeze
aur-is, -is, *f.* ear
aur-um, -ī, *n.* gold
aut . . . aut . . . either . . . or . . .
autem but
auxili-um, -ī, *n.* help
  auxili-a, -ōrum, *n.pl.* reinforcements

avār-us, -a, -um greedy, miserly
ā-vertō, -ere, -vertī, -versum I turn away, turn
  aside
av-is, -is, *f.* bird
av-us, -ī, *m.* grandfather

baline-um, -ī, *n.* bath
balneum = balineum
barbar-us, -a, -um barbarian
bell-um, -ī, *n.* war
bell-us, -a, -um pretty, charming
beāt-us, -a, -um blessed, happy
bene well
benefact-um, -ī, *n.* good deed
benefici-um, -ī, *n.* kindness
bibō, -ere, bibī I drink
bon-us, -a, -um good
  bon-a, -ōrum, *n.pl.* goods
Boreās, Boreae, *m.* the North wind
bōs, bovis, *c.* ox, bull, cow
brev-is, -e short
brevit-ās, -ātis, *f.* shortness, brevity

cadō, -ere, cecidī, cāsum I fall
caec-us, -a, -um blind; dark; hidden
caed-ēs, -is, *f.* slaughter
caedō, -ere, cecīdī, caesum I cut, beat; I kill
cael-um, -ī, *n.* sky, heaven, climate
calamit-ās, -ātis, *f.* disaster
camp-us, -ī, *m.* plain; battlefield
candid-us, -a, -um bright, white, beautiful
can-is, -is, *m.* dog
canō, -ere, cecinī, cantum I sing
capiō, -ere, cēpī, captum I capture, take, seize
captīv-us, -a, -um captive
cap-ut, -itis, *n.* head; capital
carcer, -is, *m.* prison
careō (2) + abl. I lack, am short of
carm-en, -inis, *n.* song, poem
cār-us, -a, -um dear, expensive
castell-um, -ī, *n.* fort, fortified village
castr-a, -ōrum, *n.pl.* camp
cās-us, -ūs, *m.* fall; chance
  cāsū by chance
caterv-a, -ae, *f.* crowd, company
caus-a, -ae, *f.* cause, reason; lawcase
causā + gen. for the sake of
caut-us, -a, -um cautious
caveō, -ēre, cāvī, cautum I beware (of)
cēdō, -ere, cessī, cessum I yield, give way, go
celeb-er, -ris, -e crowded, famous
celebrō (1) I celebrate
celer, -is, -e quick
cēlō (1) I hide
cēn-a, -ae, *f.* dinner
cēnō (1) I dine
cēnseō, -ēre, cēnsuī, cēnsum I vote, decide, think
cernō, -ere, crēvī, crētum I perceive, see, decide
certām-en, -inis, *n.* struggle, fight
certē certainly, at least

certō (1) I contend, fight
cert-us, -a, -um certain, sure, reliable
  certiōrem faciō I inform
  prō certō habeō I am certain
cessō (1) I linger, idle
cēter-ī, -ae, -a the rest
cib-us, -ī, *m.* food
cin-is, -eris, *m.* ash
circā (adv. & prep. + acc.) around, round about
circiter about
circum + acc. round
circum-dō, -dare, -dedī, -datum I surround
circum-stō, -stāre, -stetī I stand round, encircle
circum-veniō, -īre, -vēnī, -ventum I surround
citrā + acc. on this side of
cīvīl-is, -e civil
cīvis, -is, *c.* citizen
cīvit-ās, -ātis, *f.* state
clād-ēs, -is, *f.* disaster
clam secretly
clāmō (1) I shout
clār-us, -a, -um bright; famous
class-is, -is, *f.* fleet
claudō, -ere, clausī, clausum I shut
coepī, coepisse I begin
cōgitō (1) I think, reflect
cog-nōscō, -ere, -nōvī, -nitum I learn, get to know
cōgō, -ere, coēgī, coāctum I drive together, collect; I force, compel
cohor-s, -tis, *f.* cohort
collaudō (1) I praise
cohortor (1) I encourage, exhort
col-ligō, -ere, -lēgī, -lēctum I collect
coll-is, -is, *m.* hill
collocō (1) I place, position
colloqui-um, -ī, *n.* talk, parley
col-loquor, -ī, -locūtus I talk with
coll-um, -ī, *n.* neck
colō, -ere, coluī, cultum I till; I worship; I revere; I inhabit
color, -is, *m.* colour
columb-a, -ae, *f.* dove
com-a, -ae, *f.* hair, foliage
com-es, -itis, *c.* companion
cōm-is, -e courteous, kind
comitāt-us, -ūs, *m.* company
comiti-a, -ōrum, *n.pl.* elections
comitor (1) I accompany
commemorō (1) I mention, recount
commendāti-ō, -ōnis, *f.* recommendation
commendō (1) I recommend
com-mittō, -ere, -mīsī, -missum I entrust; I join (battle)
commoror (1) I delay
com-moveō, -ēre, -mōvī, -mōtum I move deeply, excite
commūniō (4) I fortify
commūn-is, -e common, shared
commūtāti-ō, -ōnis, *f.* change

comparō (1) I prepare, get; I compare
com-pellō, -ere, -pulī, -pulsum I drive together, I compel
complector, -ī, complexus I embrace
com-pleō, -ēre, -plēvī, -plētum I fill
complex-us, -ūs, *m.* embrace
complūr-ēs, -a several
comportō (1) I carry together
compre-hendō, -ere, -hendī, -hēnsum I seize, grasp
con-cēdō, -ere, -cessī, -cessum I retire, retreat, yield
con-cidō, -ere, -cidī I fall, collapse
conciliō (1) I unite, win over
conclāmō (1) I shout
con-currō, -ere, -currī, -cursum I run together
condemnō (1) I condemn
condici-ō, -ōnis, *f.* condition
con-dō, -ere, -didī, -ditum I hide; I store; I found
cōn-ferō, -ferre, -tulī, collātum I bring together; I compare
cōnfestim speedily, without delay
cōn-ficiō, -ere, -fēcī, -fectum I finish
cōn-fīdō, -ere, -fīsus + dat. I trust in, rely on
cōnfirmō (1) I strengthen
cōn-flīgō, -ere, -flīxī, -flīctum I fight
coniect-us, -ūs, *m.* a throw
cōn-iciō, -ere, -iēcī, -iectum I throw together; I hurl
coniugi-um, -ī, *n.* marriage
con-iūnx, -iugis, *c.* wife *or* husband
con-iungō, -ere, -iūnxī, -iūnctum I join together
  coniūnct-us, -a, -um adjoining, allied
coniūrāti-ō, -ōnis, *f.* conspiracy
coniūrāt-us, -ūs, *m.* conspirator
coniūrō (1) I conspire
conlocō = collocō
cōnor (1) I try
cōn-scendō, -ere, -scendī, -scēnsum I climb; I board
cōn-scrībō, -ere, -scrīpsī, -scrīptum I write
cōn-sequor, -ī, -secūtus I catch up, overtake
cōnservō (1) I save, preserve
cōn-sīdō, -ere, -sēdī, -sessum I sit down, settle, station myself
cōnsili-um, -ī, *n.* plan, advice
cōn-sistō, -ere, -stitī I halt, stand still
cōnsōlāti-ō, -ōnis, *f.* consolation, comfort
cōnsōlor (1) I console, comfort
cōnspect-us, -ūs, *m.* sight
cōn-spiciō, -ere, -spexī, -spectum I catch sight of, look at, observe
cōnspicor (1) I catch sight of
cōnstanti-a, -ae, *f.* constancy, steadiness
cōn-stat, -stāre, -stitit it is agreed, well-known
cōn-stituō, -ere, -stituī, -stitūtum I decide; I position
cōn-suēscō, -ere, -suēvī I am accustomed

cōnsuētūd-ō, -inis, *f.* custom

cōnsul, -is, *m.* consul

cōnsulāt-us, -ūs, *m.* consulship

con-temnō, -ere, -tempsī, -temptum I despise

contempti-ō, -ōnis, *f.* contempt

con-tendō, -ere, -tendī, -tentum I strive, march, hasten, fight

contenti-ō, -ōnis, *f.* struggle, fight, combat, effort

content-us, -a, -um content

con-tingit, -ere, -tigit it happens

continēn-s, -tis, *m.* continent

con-tineō, -ere, -tinuī, -tentum I contain, bound

continuō immediately, straight away

contrā + acc. against

con-trahō, -ere, -trāxī, -tractum I draw together, contract

con-veniō, -īre, -vēnī, -ventum I come together, gather

conventus, -ūs, *m.* gathering, meeting

con-vertō, -ere, -vertī, -versum I turn

convīci-um, -ī, *n.* insult

convīvi-um, -ī, *n.* dinner party

co-orior, -īrī, -ortus I arise

cōpi-a, -ae, *f.* plenty

    cōpi-ae, -ārum, *f.pl.* forces

cōpiōs-us, -a, -um plentiful

cor, cordis, *n.* heart

corp-us, -oris, *n.* body

cor-rigō, -ere, -rēxī, -rēctum I correct

cor-ripiō, -ere, -ripuī, -reptum I seize, snatch up

cotīdiē/cottīdiē every day

crās tomorrow

crēdō, -ere, crēdidī, crēditum + dat. I believe, trust

creō (1) I make, elect

crēscō, -ere, crēvī I increase, grow

crīm-en, -inis, *n.* charge (judicial)

crūdēl-is, -e cruel

crūdēlit-ās, -ātis. *f.* cruelty

culp-a, -ae, *f.* fault, blame

culpō (1) I blame

cum (conjunction) when; since, although

cum + abl. with

cūnctor (1) I delay

cupidit-ās, -ātis, *f.* greed, lust

cupid-us, -a, -um desirous of, eager

cupīd-ō, -inis, *f.* desire

cupiō, -ere, cupīvī, cupītum I desire

cūr? why?

cūr-a, -ae, *f.* care, anxiety

Cūri-a, -ae, *f.* senate house

cūrō (1) ut I care for; I take care that

currō, -ere, cucurrī, cursum I run

curr-us, -ūs, *m.* chariot

curs-us, -ūs, *m.* run, course, race; career

custōdi-a, -ae, *f.* custody, guard

custōdiō (4) I guard

custō-s, -dis, *m.* guard

damnō (1) I condemn

de-a, -ae, *f.* goddess

dē + abl. down from; about

dēbeō (2) I ought; I owe

dē-cēdō, -ere, -cessī, -cessum I withdraw, retire, depart from

dē-cernō, -ere, -crēvī, -crētum I decide, decree

decet (mē) it suits (me)

dē-cipiō, -ere, -cēpī, -ceptum I deceive

dēclārō (1) I make clear, declare

dec-us, -oris, *n.* glory, beauty, honour

dēdec-us, -oris, *n.* disgrace

dēdicō (1) I dedicate

dēditi-ō, -ōnis, *f.* surrender

dē-dō, -ere, -didī, -ditum I give up, surrender

dē-dūcō, -ere, -dūxī, -ductum I lead down, launch

dēfecti-ō, -ōnis, *f.* defection, revolt

dē-fendō, -ere, -fendī, -fēnsum I defend

dē-ferō, -ferre, -tulī, -lātum I carry down, report

dēfess-us, -a, -um tired

dē-ficiō, -ere, -fēcī, -fectum I fail; I revolt

dē-gredior, -ī, -gressus I go down; I come down

dē-iciō, -ere, -iēcī, -iectum I throw down

deinde then; next

dēlectō (1) I delight, please

dēl-eō, -ēre, -ēvī, -ētum I destroy

dēlici-ae, -ārum, *f.pl.* darling, sweetheart

dēligō (1) I bind fast

dē-ligō, -ere, lēgī, -lēctum I choose, pick out

dēmēn-s, -tis mad

dē-mittō, -ere, mīsī, -missum I send down, let down

dēmōnstrō (1) I show, point out

dēnique finally

dēns-us, -a, -um dense, thick

dēplōrō (1) I lament, bewail

dē-pōnō, -ere, -posuī, -positum I put down, give up

dē-prehen-dō, -ere, -dī, -sum I catch

dē-scendō, -ere, -scendī, -scēnsum I go down

dē-serō, -ere, -seruī, -sertum I desert, abandon

dē-sinō, -ere, -sīvī/-siī, situm I cease

dēsīderi-um, -ī, *n.* desire, longing

dēsīderō (1) I desire, long for, miss

dē-siliō, -īre, -siluī I jump down

dēsinō, -ere, -siī/sīvī I cease

dēsistō, -ere, -stitī I cease

dēspērāti-ō, -ōnis, *f.* desperation, hopelessness

dēspērō (1) I despair

dē-sum, -esse, -fuī + dat. I fail

dēterreō (2) I frighten off, deter

de-us, -ī, *m.* god

dēvolō (1) I fly down

dē-volvō, -ere, -volvī, -volūtum I roll down

dexter, dextera, dexterum

dexter, dextra, dextrum right

    dextrā on the right (hand)

dīcō, -ere, dīxī, dictum I say

dict-um, -ī, *n.* saying, word

dī-dūcō, -ere, -dūxī, -ductum I divide
di-ēs, -ēī, *m.* day
differō, differre, distulī, dīlātum I carry in different directions, scatter; I differ
difficil-is, -e difficult
dif-fugiō, -ere, -fūgī I flee away
dignit-ās, -ātis, *f.* worth, dignity, importance
dign-us, -a, -um + abl. worthy (of), deserving
dī-gredior, -ī, -gressus I go away
dīligēn-s, -tis careful, diligent
dīligenti-a, -ae, *f.* care, diligence
dī-ligō, -ere, -lēxī, lēctum I love, am fond of
dīmicō (1) I fight, struggle
dī-mittō, -ere, -mīsī, -missum I send away, dismiss
dīripiō, -ere, -ripuī, -reptum I plunder
dīr-us, -a, -um terrible
disciplīn-a, -ae, *f.* learning, discipline
discō, -ere, didicī I learn
discordi-a, -ae, *f.* discord
dis-cēdō, -ere, -cessī, -cessum I go away, depart
discipul-us, -ī, *m.* pupil
disert-us, -a, -um eloquent
dispers-us, -a, -um scattered
dissimil-is, -e unlike
dis-suādeō, -ēre, -suāsī, -suāsum I dissuade
dis-tribuō, -ere, -tribuī, -tribūtum I distribute
diū for a long time
dīvers-us, -a, -um contrary, different
dīv-es, -itis rich
dīviti-ae, -ārum, *f.pl.* riches
dīv-ī, -um, *m.pl.* the gods
dī-vidō, -ere, -vīsī, -vīsum I separate, divide
dīvīn-us, -a, -um divine
dō, dare, dedī, datum I give
doceō, -ere, docuī, doctum I teach
   doct-us, -a, -um learned
doleō (2) I grieve, feel pain
dolor, -is, *m.* grief, pain
dominor (1) + dat. I rule, dominate
domin-us, -ī, *m.* master, lord
domus, -ūs, *f.* home
   domī at home
dōnō (1) I give, present with
dōn-um, -ī, *n.* gift
dormiō (4) I sleep
dubitāti-ō, -ōnis, *f.* doubt
dubitō (1) I doubt, hesitate
dubi-us, -a, -um doubtful
dūcō, -ere, dūxī, ductum I lead; I marry
dulc-is, -e sweet
dum while, until
duplicō (1) I double
dūrō (1) I last, endure
dūr-us, -a, -um hard
dux, ducis, *m.* leader

ē/ex + abl. out of
ecce! look!
ē-dō, -ere, -didī, -ditum I give out

edō, ēsse, ēdī, ēsum I eat
ē-doceō, -ēre, -docuī, -doctum I teach, inform
ēducō (1) I educate
ē-dūcō, -ere, -dūxī, -ductum I lead out, lead forth
ef-ficiō, -ere, -fēcī, -fectum I accomplish, effect, cause
ef-fodiō, -ere, -fōdī, -fossum I dig up, dig out
ef-fugiō, -fugere, -fūgī I escape
ē-gredior, -ī, -gressus I go out
ēgregi-us, -a, -um remarkable, excellent, peerless
ē-lābor, -ī, -lāpsus I slip out, escape
ēloquenti-a, -ae, *f.* eloquence
ēloqui-um, -ī, *n.* the art of public speaking
emō, emere, ēmī, ēmptum I buy
enim for
eō, īre, iī, itum I go
eō (adverb) thither, to there
epistol-a/epistul-a, -ae, *f.* letter
eques, equitis, *m.* horseman
   equit-ēs, -um, *m. pl.* cavalry
equitāt-us, -ūs, *m.* cavalry
equ-us, ī, *m.* horse
ergō and so, therefore
ē-ripiō, -ere, -ripuī, -reptum I snatch away, rescue
errō (1) I wander; I am wrong
ērudiō (4) I teach
ērudīt-us, -a, -um learned
ērupti-ō, -ōnis, *f.* sally, break-out
ē-rumpō, -ere, -rūpī, -ruptum I burst out
essedāri-us, -ī, *m.* charioteer
essed-um, -ī, *n.* war chariot
ē-vādō, -ere, -vāsī I escape, get out of
etiam also; even, yet
etsī even if, although
ē-veniō, -īre, -vēnī, -ventum I turn out
ē-vertō, -ere, -vertī, -versum I overturn
ex-cēdō, -ere, -cessī I go out, depart
ex-cellō, -ere, -celuī I excel
excitō (1) I rouse
exempl-um, -ī, *n.* example
ex-eō, -īre, -iī, -itum I go out
exerceō (2) I exercise, train
exercit-us, -ūs, *m.* army
ex-hauriō, -īre, -hausī, -haustum I drain out, empty
exigu-us, -a, -um tiny
exīstimō (1) I think
exit-us, -ūs, *m.* way out, ending
ex-pellō, -ere, -pulī, -pulsum I drive out
explōrātor, -is, *m.* scout
explōrō (1) I search, investigate
ex-pōnō, -ere, -posuī, -positum I put out, disembark
ex-poscō, -ere, -poposcī I ask earnestly, demand
expugnāti-ō, -ōnis, *f.* storming, sack
expugnō (1) I take by storm
exsili-um, -ī, *n.* exile
ex-solvō, -ere, -solvī, -solūtum I loose, free, release

exspectō (1) I wait for
ex-stō, -stare, -stitī, -stitum I stand out
exsul, -is, *m.* exile
ex-tendō, -ere, -tendī, -tentum I stretch out, extend
ex-terreō, -ēre, -terruī, -territum I terrify
extrēm-us, -a, -um furthest, remote; last, final
ex-uō, -ere, -uī, -ūtum I take off, strip

fabricō (1) I make, construct
fābul-a, -ae, *f.* story
facēt-us, -a, -um witty
faci-ēs, -ēī, *f.* face, appearance
facil-is, -e easy
    facile easily
faciō, -ere, fēcī, factum I make, do
fact-um, -ī, *n.* deed
facult-ās, -ātis, *f.* opportunity
fallō, -ere, fefellī, falsum I deceive, cheat
fals-us, -a, -um false
fām-a, -ae, *f.* rumour, report, fame
fam-ēs, -is, *f.* hunger, starvation
famili-a, -ae, *f.* family, household
familiār-is, -is, *c.* friend
fās, *n.* (indecl.) right
fatīgō (1) I tire out, harass
fāt-um, -ī, *n.* fate
faveō, -ēre, fāvī, fautum + dat. I favour, support
favor, -is, *m.* favour, support
fēl-īx, -īcis lucky
fēmin-a, -ae, *f.* woman
ferē/fermē nearly, about, almost
feriō, -īre, percussī, ictum I strike, beat
ferō, ferre, tulī, lātum I carry, bring; I propose (a law)
fer-ōx, -ōcis fierce
ferr-um, -ī, *n.* iron; sword
fer-us, -a, -um wild, savage
fess-us, -a, -um tired
festīnāti-ō, -ōnis, *f.* haste
festīnō (1) I hasten
fid-ēs, -eī, *f.* loyalty, trust, faith, promise
fidēl-is, -e loyal
fīli-a, -ae, *f.* daughter
fīliol-a, -ae, *f.* little daughter
fīli-us, -ī, *m.* son
fingō, -ere, fīnxī, fictum I make up, invent
fīniō (4) I limit, bound
fīn-is, -is, *m.* end
    fīn-ēs, -ium, *m. pl.* bounds; territory, country
fīnitim-us, -a, -um neighbouring
fīō, fierī, factus I become, am made
firm-us, -a, -um strong, firm, reliable
flamm-a, -ae, *f.* flame
fleō, -ēre, flēvī, flētum I weep
flēt-us, -ūs, *m.* weeping
flōrēns, florentis flourishing
flōs, flōris, *m.* flower
fluct-us, -ūs, *m.* wave
flūm-en, -inis, *n.* river

flu-ō, -ere, -xī I flow
fluvi-us, -ī, *m.* river
foed-us, -eris, *n.* treaty; agreement, pact
fōrm-a, -ae, *f.* shape, beauty
fōrmōs-us, -a, -um beautiful
fortasse perhaps
forte by chance
fort-is, -e brave, strong
fortūn-a, -ae, *f.* fortune
for-um, -ī, *n.* forum; public life
foss-a, -ae, *f.* ditch
foveō, -ēre, fōvī, fōtum I nourish, nurture, warm, foster, respect
frangō, -ere, frēgī, frāctum I break
frāt-er, -ris, *m.* brother
fraus, fraudis, *f.* deceit, cheating
frequē-ns, -ntis crowded, in crowds
frequenti-a, -ae, *f.* crowd, throng
frīgid-us, -a, -um cold
frīg-us, -oris, *n.* cold
frōns, frontis, *f.* forehead; front
frōns, frondis, *f.* foliage, leaves
frūment-um, -ī, *n.* corn
fruor, fruī, frūctus + abl. I enjoy
frūstrā in vain
fug-a, -ae, *f.* flight
fugiō, -ere, fūgī, fugitum I flee
fūm-us, -ī, *m.* smoke
fundō, -ere, fūdī, fūsum I pour; I rout, cause to flee
fund-us, -ī, *m.* estate, farm
fūn-us, -eris, *n.* funeral, death
furō, -ere I rage, rave, am mad
furor, -is, *m.* fury, madness
fūror (1) I steal
fūrtim furtively, secretly

gaudeō, -ēre, gāvīsus I rejoice, rejoice in
gaudi-um, -ī, *n.* joy
gemit-us, -ūs, *m.* groan, groaning
gemō, -ere, gemuī I groan
gēns, gentis, *f.* race, people
gen-ū, -ūs, *n.* knee
gen-us, -eris, *n.* race, family; kind
gerō, -ere, gessī, gestum I carry, wear; I do; I wage
    mē gerō I behave
glaci-ēs, -ēī, *f.* ice
gladi-us, -ī, *m.* sword
gradior, -ī, gressus I step, walk, go
grad-us, -ūs, *m.* step
Graec-us, -a, -um Greek
grāti-a, -ae, *f.* thanks, favour, influence
    grātiās agō I thank
grātulāti-ō, -ōnis, *f.* congratulations
grātulor (1) + dat. I congratulate
grāt-us, -a, -um pleasing, thankful
grav-is, -e heavy; serious
gravit-ās, -ātis, *f.* weight; seriousness
gubernātor, -is, *m.* helmsman

habeō (2) I have; I consider
habitō (1) I live in, inhabit
haereō, -ēre, haesī, haesum I stick
hast-a, -ae, *f.* spear
haud not
hauriō, -īre, hausī, haustum I drain, drink down
hībern-us, -a, -um (of) winter
  hībern-a, -ōrum, *n. pl.* winter quarters
hīc here
hiems, hiemis, *f.* winter
hinc hence, from here
hodiē today
hom-ō, -inis, *c.* man, human being
honest-us, -a, -um honourable
honor, -is, *m.* honour; public office
hōr-a, -ae, *f.* hour
horreō, -ēre, horruī I bristle, shudder at, stand in
  awe of
horrid-us, -a, -um horrid
hortor (1) I encourage
hort-us, -ī, *m.* garden
hosp-es, -itis, *c.* guest, host; stranger
host-is, -is, *c.* enemy
hūc hither, to here
hūmān-us, -a, -um humane, civilized
hum-us, -ī, *f.* ground
  humī on the ground

iaceō (2) I lie
iaciō, -ere, iēcī, iactum I throw
iacul-um, -ī, *n.* spear, javelin
iam now, already
iamdūdum now for a long time
ibi there
identidem again and again
idōne-us, -a, -um suitable
igitur therefore, and so
ignār-us, -a, -um ignorant (of)
ignāv-us, -a, -um cowardly, lazy
ign-is, -is, *m.* fire
ignōrō (1) I am ignorant, do not know
ig-nōscō, -ere, nōvī + dat. I pardon
ignōt-us, -a, -um unknown
illīc there
illinc from there
illūc to that place
illūstr-is, -e bright, famous
imbell-is, -e unwarlike
immān-is, -e monstrous, inhuman, huge
immātūr-us, -a, -um untimely
immem-or, -oris forgetful
immineō, -ēre I hang over, project over;
  I threaten
immortāl-is, -e immortal
impedīment-a, -ōrum, *n. pl.* baggage
impediō (4) I hinder
impendeō, -ēre I hang over; I threaten
imperātor, -is, *m.* general; emperor
imperāt-um, -ī, *n.* order

imperi-um, -ī, *n.* order, command, empire
imperō (1) + dat. I order
impet-us, -ūs, *m.* attack; violence, dash
impi-us, -a, -um impious; undutiful
improb-us, -a, -um wicked
imprūdēn-s, -tis imprudent, silly
imprūdenti-a, -ae, *f.* imprudence, folly
impudenti-a, -ae, *f.* shamelessness
inān-is, -e empty
in-cēdō, -ere, -cessī, -cessum I go, advance
incendi-um, -ī, *n.* fire
in-cendō, -ere, -cendī, -cēnsum I set on fire, burn
incert-us, -a, -um uncertain, doubtful
in-cidō, -ere, -cidī I fall into, light on
incipiō, -ere, -cēpī, -ceptum I begin
incitō (1) I urge on
inclūdō, -ere, -clūsī, -clūsum I shut in
incognit-us, -a, -um unknown
in-colō, -ere, -coluī, -cultum I live in, inhabit
incolum-is, -e safe
incult-us, -a, -um uncultivated; barbarous
inde from there; then
indignor (1) I am indignant
indign-us, -a, -um unworthy (of); undeserved
in-dūcō, -dūcere, -dūxī, -ductum I bring into
industri-a, -ae, *f.* industry, hard work
in-eō, -īre, -iī, -itum I enter; I begin
inerm-is, -e unarmed, defenceless
īnfēl-īx, -īcis unlucky
īn-ferō, -ferre, -tulī, illātum I bring into, carry
  against
īnfest-us, -a, -um hostile, dangerous; endangered,
  disordered
īnfidēlit-ās, -ātis, *f.* infidelity, treachery
īnflammō (1) I set on fire, burn
īnfirm-us, -a, -um weak
ingeni-um, -ī, *n.* talents, character
ingēn-s, -tis huge
ingrāt-us, -a, -um ungrateful
in-gredior, -gredī, -gressus I go into, enter
īn-iciō, -ere, iecī, -iectum I throw into
inimīc-us, -ī, *m.* enemy
inīqu-us, -a, -um unequal, unjust, unfavourable
initi-um, -ī, *n.* beginning
iniūri-a, -ae, *f.* injury, wrong
inopi-a, -ae, *f.* scarcity, shortage
inquam, inquit, inquiunt I, he, they say/said
īnsci-us, -a, -um not knowing, ignorant
īn-sequor, -ī, -secūtus I pursue; I follow after
īn-sideō, -ēre, -sēdī, -sessum I sit upon, occupy
īnsidi-ae, -ārum, *f. pl.* ambush, plot, treachery
īn-sīdō, -ere, -sēdī, -sessum I settle down, sit
  upon, occupy
īnsign-is, -e distinguished
īnspectō (1) I look at, observe
īn-spiciō, -ere, -spexī, -spectum I look at, observe
īn-struō, -ere, -strūxī, -structum I draw up
īnsul-a, -ae, *f.* island
īn-sum, -esse, -fuī I am in, among

intel-legō, -ere, -lēxī, -lēctum I understand
in-tendō, -ere, -tendī, -tentum I intend, aim
intent-us, -a, -um eager
inter + acc. among, between
inter-cēdō, -cēdere, -cessī, -cessum I come between
inter-cipiō, -ere, -cēpī, -ceptum I intercept
interdiū in the day time
intereā meanwhile
inter-eō, -īre, -iī I perish, die
inter-ficiō, -ere, -fēcī, -fectum I kill
interim meanwhile
inter-mittō, -ere, -mīsī, -missum I leave off, let pass
interrogō (1) I question, ask
inter-sum, -esse, -fuī I take part in
intrā + acc. inside
intrō (1) I enter
intueor (2) I gaze at
intus (adverb) within
inūtil-is, -e useless
in-vādō, -ere, -vāsī, -vāsum I attack
in-vehor, -ī, -vectus (in) I attack (verbally)
in-veniō, īre, -vēnī, -ventum I find
in-videō, -ēre, -vīdī, -vīsum + dat. I am jealous of, envy
invidi-a, -ae, f. envy, malice
in-vīsō, -ere, -vīsī I visit
invīt-us, -a, -um unwilling
invi-us, -a, -um impassable
ips-e, -a, -um himself, herself, itself
īr-a, -ae, f. anger
īrāscor, -ī, īrātus I am angry
īrāt-us, -a, -um angry
ist-e, -a, -ud that
istīc there
ita thus, so
itaque and so
iter, itineris, n. journey, march
iterum again; a second time
iubeō, -ēre, iussī, iussum I order
iūcund-us, -a, -um pleasant, delightful
iūd-ex, -icis, c. judge; juryman
iūdici-um, -ī, n. judgement; law-court, trial
iūdicō (1) I judge
iug-um, -ī, n. yoke; mountain pass, ridge
iūment-um, -ī, n. beast of burden
iungō, -ere, iūnxī, iūnctum I join
iūs, iūris, n. right, law
iussū by order
iūstiti-a, -ae, f. justice
iūst-us, -a, -um just
iuvenīl-is, -e youthful
iuven-is, -is, m. young man
iuvō, -āre, iūvī, iūtum I help; I please

lābor, -ī, lāpsus I slip
labor, -is, m. labour, toil
labōriōs-us, -a, -um laborious
labōrō (1) I work, suffer

lacrim-a, -ae, f. tear
lac-us, -ūs, m. lake
laedō, -ere, laesī, laesum I hurt
laetiti-a, -ae, f. joy, happiness
laet-us, -a, -um joyful, happy
lap-is, -idis, m. stone
lāps-us, -ūs, m. falling, slide
lateō (2) I lie hidden
lat-us, -eris, n. side
lāt-us, -a, -um wide
laudō (1) I praise
laus, laudis, f. praise
lavō, -āre, lāvī, lautum I wash
lectīc-a, -ae, f. litter
lect-us, -ī, m. couch, bed
lēgāt-us, -ī, m. deputy, ambassador
lēgātus legiōnis legionary commander
legi-ō, -ōnis, f. legion
legō, -ere, lēgī, lēctum I choose; I read
leō, leōnis, m. lion
lev-is, -e light
levō (1) I lighten, relieve
lēx, lēgis, f. law
liber, librī, m. book
līber, lībera, līberum free
līberāl-is, -e gracious, generous
līber-ī, -ōrum, m. pl. children
līberō (1) I free
lībert-ās, -ātis, f. freedom
licet (mihi) it is lawful for me; I am allowed
līm-en, -inis, n. threshold
lingu-a, -ae, f. tongue, language
linquō, -ere, līquī I leave
lītorāl-is, -e of the shore
litter-a, -ae, f. letter (of the alphabet)
litter-ae, -ārum, f. pl. a letter, literature
līt-us, -oris, n. shore
locupl-ēs, -ētis rich
loc-us, -ī, m. place
longē far off, at a distance; by far
longinqu-us, -a, -um far off, distant
long-us, -a, -um long
loquor, -ī, locūtus I speak, say
lūct-us, -ūs, m. grief
lūdō, -ere, lūsī, lūsum I play
lūd-us, -ī, m. school; play, game
lūgeō, -ēre, lūxī, luctum I grieve, mourn
lūm-en, -inis, n. light
lūn-a, -ae, f. moon
lup-us, -ī, m. wolf
lūsor, -is, m. playful poet
lūx, lūcis, f. light

maeror, -is, m. grief
maest-us, -a, -um sad
magis more
magist-er, -rī, m. master
magistrāt-us, -ūs, m. magistate
magnitūd-ō, -inis, f. size
magnopere greatly

magn-us, -a, -um great
māiōr-ēs, -um, *m. pl.* ancestors
maledict-um, -ī, *n.* abuse
mālō, mālle, māluī I prefer
mal-us, -a, -um evil, bad
mal-a, -ōrum, *n. pl.* evils, troubles
mandāt-a, -ōrum, *n. pl.* orders
mandō(1) I instruct
māne (adverb) early
maneō, -ere, mānsī, mānsum I wait, wait for
man-us, -ūs, *f.* hand, band
mar-e, -is, n. sea
maritim-us, -a, -um sea
māt-er, -ris, *f.* mother
mātern-us, -a, -um maternal
mātrōn-a, -ae, *f.* married woman, wife
mātūrē in good time
mātūr-us, -a, -um timely, early
medic-us, -ī, *m.* doctor
medi-us, -a, -um middle
membr-um, -ī, *n.* limb
meminī, meminisse I remember
memor, -is mindful of, remembering, recalling
memorābil-is, -e memorable, remarkable
memori-a, -ae, *f.* memory, recollection
memorō (1) I mention, narrate
mēns, mentis, *f.* mind
mēns-is, -is, *m.* month
mercātor, -is, *m.* merchant
mereō/mereor (2) I deserve, merit; I serve
merit-um, -ī, *n.* desert, reward
merīdi-ēs, -ēī, *m.* midday
merīdiān-us, -a, -um of midday
met-uō, -ere, -uī I fear
met-us, -ūs, *m.* fear
me-us, -a, -um my
migrō (1) I remove, depart
mīles, -itis, *m.* soldier
mīliēns a thousand times
mīlitār-is, -e military
mīliti-a, -ae, *f.* military service
mīlitō (1) I serve as a soldier, campaign
min-a, -ae, *f.* threat
min-āx, -ācis threatening
minor (1) I threaten
mīrand-us, -a, -um wonderful
mīror (1) I wonder at, admire
mīr-us, -a, -um wonderful
miser, -a, -um unhappy, wretched
misereor, -ērī, miseritus I pity
miseri-a, -ae, *f.* misery
mittō, -ere, mīsī, missum I send
modo (adverb) only, recently
modo. . . modo. . . now . . . now . . .
mod-us, -ī, *m.* way; limit
moeni-a, -um, *n. pl.* walls, fortifications
mōl-ēs, -is, *f.* heavy mass, difficulty, labour
moll-is, -e soft
mōment-um, -ī, *n.* importance; moment

moneō (2) I warn, advise
moniment-um, -ī, *n.* monument, memorial
mōns, montis, *m.* mountain
mōnstrō (1) I show
montān-us, -a, -um mountain
mor-a, -ae, *f.* delay
morb-us, -ī, *m.* disease
morior, morī, mortuus I die
moror (1) I delay
mors, mortis, *f.* death
mortāl-is, -e mortal
mōs, mōris, *m.* custom
moveō, -ēre, mōvī, mōtum I move
mox soon
mulier, -is, *f.* woman
multitūdō, -inis, *f.* multitude, crowd
mult-us, -a, -um much, many
mūn-us, -eris, *n.* gift; duty
mūniō (4) I fortify
mūnīti-ō, -ōnis, *f.* fortification
mūr-us, -ī, *m.* wall

nact-us, -a, -um having obtained
nam for
nārrō (1) I narrate, tell
nāscor, -ī, nātus I am born
nātāl-is, -is, *m.* birthday
nāti-ō, -ōnis, *f.* tribe
nātūr-a, -ae, *f.* nature
nāt-us, -a, -um born, old
naut-a, -ae, *m.* sailor
nāvāl-is, -e naval
nāvigō (1) I sail
nāv-is, -is, *f.* ship
nē lest
nē . . . quidem not even
necesse (indecl.) necessary, inevitable
necessāri-us, -a, -um necessary
necō (1) I kill, butcher
nefāri-us, -a, -um wicked
neg-legō, -ere, -lēxī, -lēctum I neglect
negō (1) I deny
negōti-um, -ī, *n.* business
nēmō, nēminis no one
nem-us, -oris, *n.* grove
nēquīquam in vain
nesciō (4) I do not know
nesci-us, -a, -um unknowing, unaware
nī = nisi
nihil (indecl.) nothing
nihilōminus nevertheless
nīl (indecl.) nothing
nimis too much
nisi unless; except
niteō, -ēre, nituī I shine
nix, nivis, *f.* snow
nōbil-is, -e noble, famous
noceō, -ēre, nocuī + dat. I harm
noctū by night
nocturn-us, -a, -um of the night

nōlō, nōlle, nōluī I am unwilling, do not wish
nōm-en, -inis, *n.* name
nōndum not yet
nōnnūll-ī, -ae, -a some
nōnnunquam sometimes
nōscō, -ere, nōvī, nōtum I get to know, learn
nost-er, -ra, -rum our
nōt-us, -a, -um known
novit-ās, -ātis, *f.* novelty, newness
nov-us, -a, -um new
nox, noctis, *f.* night
nūb-ēs, -is, *f.* cloud
nūbō, -ere, nūpsī, nuptum + dat. I marry
nūd-us, -a, -um bare, nude
nūg-ae, -ārum, *f. pl.* trifles, nonsense
nūll-us, -a, -um no
numer-us, -ī, *m.* number, metre
numquam never
nunc now
nūntiō (1) I announce
nūnti-us, -ī, *m.* message, messenger
nūper lately
nusquam nowhere

ob + acc. on account of
obit-us, -ūs, *m.* death
oblīvīscor, -ī, oblītus + gen. I forget
obscūr-us, -a, -um dark, dim
ob-ses, -sidis, *c.* hostage
ob-sideō, -ēre, -sēdī, -sessum I besiege
obsidi-ō, -ōnis, *f.* siege
ob-stō, -stāre, -stitī I stand in the way of; I block
ob-tineō, -ēre, -tinuī, -tentum I hold, possess, occupy, keep
obviam eō + dat. I go to meet
obvi-us, -a, -um meeting, to meet
occāsi-ō, -ōnis, *f.* opportunity
occās-us, -ūs, *m.* (sōlis) sunset
occupō (1) I seize, occupy
oc-cīdō, -ere, -cīdī, -cīsum I beat; I kill
oc-cidō, -ere, -cidī I fall; I set; I die
occultō (1) I hide
occult-us, -a, -um hidden, secret
occupō (1) I seize
oc-currō, -ere, -currī + dat. I meet
ocell-us, -ī, *m.* (little) eye
ocul-us, -ī, *m.* eye
ōdī, ōdisse I hate
odi-um, -ī, *n.* hatred
offerō, -ferre, obtulī, oblātum I offer
offici-um, -ī, *n.* duty
ōlim once; at some time
omnīnō altogether; at all
omn-is, -e all
onerāria (nāvis) merchant ship, transport vessel
on-us, -eris, *n.* burden
oper-a, -ae, *f.* member of a gang
op-ēs, -um, *f. pl.* riches, wealth; resources
opīni-ō, -ōnis, *f.* opinion
opīnor (1) I think

oppid-um, -ī, *n.* town
opportūn-us, -a, -um opportune, at the right time
oppugnāti-ō, -ōnis, *f.* attack
oppugnō (1) I attack
ops, opis, *f.* aid, help
opēs, opum, *f. pl.* wealth, resources
opus, operis, *n.* work; siege work
opus est mihi + abl. I have need of
optō (1) I wish for, pray for
ōrāti-ō, -ōnis, *f.* speech
ōrātor, -is, *m.* speaker
orb-is, -is, *m.* circle
orbis terrārum the world
ōrd-ō, -inis, *m.* rank; class
orīg-ō, -inis, *f.* origin
orior, orīrī, ortus I arise
ōrnō (1) I adorn, equip
ōrō (1) I beg, pray
ōs, ōris, *n.* mouth, face
os, ossis, *n.* bone
os-tendō, -ere, -tendī, -tentum I show
ōtiōsus, -a, -um at leisure
ōti-um, -ī, *n.* leisure

palam openly
pal-ūs, -ūdis, *f.* marsh
pār, paris equal
parcō, -ere, pepercī + dat. I spare
parēn-s, -tis, *c.* parent
pāreō + dat. I obey
parō (1) I prepare; I acquire
pars, partis, *f.* part
part-ēs, -ium, *f. pl.* party
parv-us, -a, -um small
pass-us, -ūs, *m.* pace
   mille passūs one mile
pāstor, -is, *m.* shepherd
pate-faciō, -ere, -fēcī, -factum I open
pateō (2) I lie open
pat-er, -ris, *m.* father
patior, -ī, passus I suffer; I allow
patri-a, -ae, *f.* native land
patrōn-us, -ī, *m.* patron
pauc-ī, -ae, -a few
paucit-ās, -ātis, *f.* fewness, scarcity
paulātim little by little
paulīsper for a little time
paulum (adverb) a little
paulō (by) a little
pauper, -is poor
paveō, -ēre I tremble, am afraid
pāx, pācis, *f.* peace
pect-us, -oris, *n.* breast
pecūni-a, -ae, *f.* money
pec-us, -oris, *n.* herd, flock, cattle
peditāt-us, -ūs, *m.* infantry
pedit-ēs, -um, *m. pl.* infantry
pellō, -ere, pepulī, pulsum I drive
pendeō, -ēre, pependī I hang
pendō, -ere, pependī, pēnsum I weigh; I pay

penetrō (1) I penetrate
per + acc. through
per-dō, -ere, -didī, -ditum I lose, waste
per-eō, -īre, -iī, -itum I perish, die
perficiō, -ere, -fēcī, -fectum I complete
perfidi-a, -ae, *f.* treachery
perfug-a, -ae, *m.* deserter
per-fugiō, -ere, -fūgī, -fugitum I flee (for refuge)
perīcul-um, -ī, *n.* danger
perīt-us, -a, -um skilled
per-legō, -ere, -lēgī, -lēctum I read through
per-maneō, -ēre, -mānsī, -mānsum I remain,
  persist
per-mittō, -ere, -mīsī, -missum I permit, allow
perpetu-us, -a, -um perpetual
  in perpetuum for ever
per-rumpō, -ere, -rūpī, -ruptum I break through
per-sequor, -ī, -secūtus I pursue
per-scrībō, -ere, -scrīpsī, -scrīptum I write in full
per-spiciō, -ere, -spexī, -spectum I look at,
  perceive
per-suādeō, -ēre, -suāsī, -suāsum + dat. I
  persuade
perterreō (2) I terrify
perturbāti-ō, -ōnis, *f.* confusion
perturbō (1) I throw into confusion
pēs, pedis, *m.* foot
petō, -ere, petīvī, petītum I seek, ask, attack
piet-ās, -ātis, *f.* loyalty, devotion, piety
pisc-is, -is, *m.* fish
pi-us, -a, -um loyal, pious
placeō (2) + dat. I please
placet (2) **mihi** it pleases me; I decide
plānē plainly, clearly
plān-us, -a, -um flat
plaus-us, -ūs, *m.* applause
plēbs, plēbis, *f.* common people
plēn-us, -a, -um full, abundant
plērīque, plēraeque, plēraque most
plūs, plūris more
poēm-a, -atis, *n.* poem
poen-a, -ae, *f.* punishment, penalty
poenās dō I pay the penalty
Poen-ī, -ōrum, *m. pl.* Carthaginians
poēt-a, -ae, *m.* poet
polliceor (2) I promise
pond-us, -eris, *n.* weight
pōnō, -ere, posuī, positum I place, put
pōns, pontis, *m.* bridge
populār-is, -e popular
populār-ēs, -ium, *m. pl.* the popular party
popul-us, -ī, *m.* people
port-a, -ae, *f.* gate
portic-us, -ūs, *f.* portico, colonnade
portō (1) I carry
port-us, -ūs, *m.* harbour
pos-sideō, -ēre, -sēdī, -sessum I possess, have
poscō, -ere, poposcī I demand
possum, posse, potuī I am able; I can

post + acc. after, behind
posteā afterwards
poster-us, -a, -um the next
posthāc after this
postquam (conjunction) after
postrīdiē the next day
postulō (1) I demand
potēn-s, -tis powerful
potenti-a, -ae, *f.* power
potest-ās, -ātis, *f.* power
potissimum most
potius rather
praebeō (2) I offer, provide, show
prae-ceps, -cipitis headlong, precipitous, hasty
praeceptor, -is, *m.* teacher
prae-cipiō, -ere, -cēpī, -ceptum I teach
praeclār-us, -a, -um famous
praed-a, -ae, *f.* booty
praed-ō, -ōnis, *m.* pirate
prae-ficiō, -ere, -fēcī, -fectum I put in charge of
praemi-um, -ī, *n.* reward
prae-mittō, -ere, -mīsī, -missum I send ahead
praesertim especially
praesidi-um, -ī, *n.* garrison
prae-stō, -āre, -stitī, -stitum I excel; I show
prae-sum, -esse, -fuī + dat. I am in charge of
praeter + acc. past, beyond, except for, beside,
  along
praetereā moreover
prater-eō, -īre, -iī, -itum I pass by, overtake
precor (1) I pray
prec-ēs, -um, *f. pl.* prayers
premō, -ere, pressī, pressum I press, overwhelm; I
  darken
preti-um, -ī, *n.* price
prīdiē the day before
prīm-us, -a, -um first
  prīmum (adverb) first
  in prīmīs especially
prīn-ceps, -cipis, *m.* chief, leading man; first
prīncipi-um, -ī, *n.* beginning
prior, prius former
prīstin-us, -a, -um former
priusquam before
prīvāt-us, -a, -um private
prō + abl. in front of, on behalf of, instead of, in
  return for
prō-cēdō, -ere, -cessī, -cessum I go forward,
  proceed, advance
procul far, at a distance
prō-currō, -ere, -currī, -cursum I run forward
prō-dō, -ere, -didī, -ditum I betray
proeli-um, -ī, *n.* battle
prō-ficiō, -ere, fēcī, -fectum I make progress
pro-ficīscor, -ī, -fectus I set out
prō-gredior, -ī, -gressus I advance
prohibeō (2) I prevent
prō-iciō, -ere, -iēcī, -iectum I throw down
prō-lābor, -ī, -lāpsus I fall down, slip forward

prōmiss-um, -ī, *n.* promise
prō-mittō, -ere, -mīsī, -missum I promise
prope + acc. near
properō(1) I hasten
propinqu-us, -a, -um near; kinsman
prō-pōnō, -ere, -posuī, -positum I put forward, propose
propter + acc. on account of
prō-sequor, -ī, -secūtus I escort
prōtinus (adverb) straight away
prosperē successfully
prōsum, prōdesse, prōfuī + dat. I benefit, help
prō-vehō, -ere, -vēxī, -vectum I carry forward
prō-videō, -ēre, -vīdī, -vīsum I foresee, provide
prōvinci-a, -ae, *f.* province
proxim-us, -a, -um nearest, next
prūdēn-s, -tis sensible, wise
prūdenti-a, -ae, *f.* good sense
pudet (mē) I am ashamed
pudor, -is, *m.* shame
puell-a, -ae, *f.* girl
puer, -ī, *m.* boy
pueriti-a, -ae, *f.* boyhood
pugn-a, -ae, *f.* battle
pugnō (1) I fight
pulcher, pulchra, pulchrum beautiful
pulchritūd-ō, -inis, *f.* beauty
pulv-is, -eris, *m.* dust
pūniō (4) I punish
pupp-is, -is, *f.* stern, ship
pūriter purely
pūr-us, -a, -um pure
putō (1) I think, consider

quā where
quaerō, -ere, quaesīvī, quaesītum I seek, ask
quāl-is, -e what sort of? what?
quam than; how; as
quamquam although
quandō? when?
quant-us, -a, -um? how great?
quārē and so
quasi as if
querēl-a, -ae, *f.* complaint
queror, -ī, questus I complain
quia because
quīcumque, quaecumque, quodcumque whoever, whatever
quīdam, quaedam, quoddam a, a certain
quidem indeed
qui-ēs, -ētis, *f.* rest
quis? quid? who? what?
quis, quid anyone, anything
quisquam, quicquam anyone, anything
quisque, quaeque, quodque each
quisquis, quidquid whoever
quō? whither? where to?
quod because
quōmodo? how?
quondam once, formerly; at some future time

quoniam since
quoque also
quot? (indecl. adj.) how many? as many
quotiēns how often

rapiō, -ere, rapuī, raptum I snatch
rārō (adverb) rarely
rār-us, -a, -um rare, scanty
rati-ō, -ōnis, *f.* reason, plan
rat-is, -is, *f.* raft, ship
rebelli-ō, -ōnis, *f.* rebellion
re-cēdō, -ere, cessī, -cessum I retire, retreat, go away
rec-ēns, -entis recent
re-cipiō, -ere, -cēpī, -ceptum I take back
   mē recipiō I retire, retreat
recitō (1) I read aloud
reconciliō (1) I reconcile
recordor (1) I remember
recreō (1) I restore, revive
red-eō, -īre, -iī, -itum I return
redit-us, -ūs, *m.* return
referō, referre, rettulī, relātum I carry back; I report
   pedem referō I retreat
re-ficiō, -ere, -fēcī, -fectum I repair
rēgīn-a, -ae, *f.* queen
regi-ō, -ōnis, *f.* region, district
rēgi-us, -a, -um royal
rēgn-um, -ī, *n.* kingdom, power
rēgnō (1) I rule as king
regō, -ere, rēxī, rēctum I rule
re-gredior, -ī, -gressus I return
religi-ō, -ōnis, *f.* religion
re-linquō, -ere, -līquī, -lictum I leave behind
reliqu-us, -a, -um remaining, the rest
rēmigō (1) I row
rēm-us, -ī, *m.* oar
renovō (1) I renew
reor, rērī, ratus I think
repellō, -ere, reppulī, repulsum I drive back, I reject
repente suddenly
repentīn-us, -a, -um sudden
reperiō, -īre, repperī, repertum I find
re-petō, -ere, -petiī, -petītum I seek again
re-pleō, -ēre, -plēvī, -plētum I fill up, replenish
re-pōnō, -ere, -posuī, -positum I put back
re-portō (1) I carry back
re-poscō, -ere I demand back
re-prehendō, -ere, -prehendī, -prehēnsum I blame
re-quiēscō, -ere, -quiēvī I rest
re-quīrō, -ere, -quīsiī, -quīsītum I look for, ask for
rēs, reī, *f.* thing, matter
re-sistō, -ere, -stitī, -stitum + dat. I resist
resonō (1) I resound, echo
respectō (1) I look back
re-spiciō, -ere, -spexī, -spectum I look back
re-spondeō, -ēre, -spondī, -spōnsum I reply, answer

rēspūblica, rēīpūblicae, *f.* the republic, public affairs

re-stituō, -ere, -stituī, -stitūtum I restore, put back

re-vertor, -ī, -versus I return

re-vellō, -ere, -vellī, -vulsum I pluck out, undo

re-vīsō, -ere, -vīsī I revisit

rēx, rēgis, *m.* king

rhētoric-us, -a, -um rhetorical

rīdeō, -ēre, rīsī, rīsum I laugh (at)

rīp-a, -ae, *f.* bank

rīv-us, -ī, *m.* stream

rix-a, -ae, *f.* quarrel

rōb-ur, -oris strength, flower

rogō (1) I ask, I ask for

rōstra, -ōrum, *n.pl.* speaker's platform

rūmor, -is, *m.* rumour, report

rumpō, -ere, rūpī, ruptum I break

ruō, -ere, ruī I rush

rūp-ēs, -is, *f.* crag, rock, cliff

rūrsus (adverb) again

rūs, rūris, *n.* countryside

sacer, sacra, sacrum sacred; cursed

sacerd-ōs, -ōtis, *m.* priest

sacrificō (1) I sacrifice

saepe often

saev-us, -a, -um savage

sagitt-a, -ae, *f.* arrow

sal-ūs, -ūtis, *f.* health, safety, greetings

salūtō (1) I greet

salv-us, -a, -um safe

sānct-us, -a, -um holy

sangu-is, -inis, *m.* blood

sapiēn-s, -tis wise

sapienti-a, -ae, *f.* wisdom

satis enough

sauci-us, -a, -um wounded

sax-um, -ī, *n.* rock

scelus, -eris, *n.* crime

scienti-a, -ae, *f.* knowledge

scindō, -ere, scidī, scissum I cut, break

sciō (4) I know

scrīb-a, -ae, *m.* secretary

scrībō, -ere, scrīpsī, scrīptum I write

scūt-um, -ī, *n.* shield

sē-cēdō, -ere, -cessī, -cessum I retire

secō, -āre, secuī, sectum I cut

secund-us, -a, -um second, favourable

sedeō, -ēre, sēdī, sessum I sit

sēd-ēs, -is, *f.* seat

semel once

sēmimortu-us, -a, -um half-dead

semper always

senex, senis, *m.* old man

senior, -is older

sēns-us, -ūs, *m.* feeling

sententi-a, -ae, *f.* opinion, vote

sentiō, -īre, sēnsī, sēnsum I feel, perceive

sequor, -ī, secūtus I follow

serm-ō, -ōnis, *m.* conversation, talk, speech

sērō late

servō (1) I save, preserve; I observe

serv-us, -ī, *m.* slave

sēsterti-us, -ī, *m.* sesterce

seu . . . seu . . . whether . . . or . . .

sevēr-us, -a, -um severe

sī if

sīc thus

sīcut just as, just as if

sīd-us, -eris, *n.* constellation, star

significō (1) I signify

sign-um, -ī, *n.* sign, standard, signal

silenti-um, -ī, *n.* silence

silv-a, -ae, *f.* wood, forest

silvestr-is, -e woodland, silvan

simil-is, -e like

simul together

simul ac/atque at the same time as

simulāti-ō, -ōnis, *f.* pretence

simulō (1) I pretend

sīn but if

sine + abl. without

sinister, sinistra, sinistrum left

sinistrā on the left

sinō, sinere, sīvī, situm I allow

sistō, -ere, stitī, statum I stand

sit-us, -a, -um sited

sīve . . . sīve . . . whether . . . or . . .

soci-us, -ī, *m.* ally, partner

sōl, sōlis, *m.* sun

soleō, -ēre, solitus I am accustomed

sōlitūdō, -inis, *f.* solitude, loneliness

sōlus, -a, -um alone

sōlum (adverb) only

solvō, -ere, solvī, solūtum I loose, cast off

somn-us, -ī, *m.* sleep

sonō, -āre, sonuī I sound, make a noise

son-us, -ī, *m.* sound

sordid-us, -a, -um dirty, mean

soror, -ōris, *f.* sister

sors, sortis, *f.* chance, lot

spati-um, -ī, *n.* space (of time or distance)

speci-ēs, -ēī, *f.* appearance; pretence

speciō, -ere, spexī I look at

spectācul-um, -ī, *n.* sight, show

spectātor, -is, *m.* spectator

spectō (1) I watch

spernō, -ere, sprēvī, sprētum I despise

spērō (1) I hope, expect

spēs, spēī, *f.* hope

sponte (meā) of (my) own accord

statim at once

stati-ō, -ōnis, *f.* post

stat-us, -ūs, *m.* state (of affairs)

stīpō (1) I press, crowd together; I surround

stirp-s, -is, *f.* stock, family

stō, stāre, stetī, statum I stand

strēnu-us, -a, -um strenuous, active, vigorous

strepit-us, -ūs, *m.* noise

studeō (2) + dat. I am keen on
studiōs-us, -a, -um eager, studious
studi-um, -ī, n. eagerness, effort, study
stupeō (2) I am amazed, am stunned
suādeō, -ēre, suāsī, suāsum + dat. I persuade
sub + abl. under
sub-dūcō, -ere, -dūxī, -ductum I lead down; I
  beach; I remove, steal
sub-eō, -īre, -iī I approach; come into one's mind
sub-iciō, -ere, -iēcī, -iectum I throw under
subiect-us, -a, -um lying under
subigō, -ere, -ēgī, -āctum I subdue
subitō (adverb) suddenly
sublātum – see tollō
sub-sequor, -ī, -secūtus I follow closely
sub-sīdō, -ere, -sēdī, -sessum I sit down, sink
sub-veniō, -īre, -vēnī, -ventum + dative I come to
  help
success-us, -ūs, m. success
summ-a, -ae, f. sum, main point
summ-us, -a, -um highest, greatest
sūmō, -ere, sūmpsī, sumptum I take, take up
super + acc. above
superb-us, -a, -um proud
superi-or, -us higher, earlier, superior
superō (1) I overcome
supplici-um, -ī, n. punishment, death penalty
surgō, -ere, surrēxī, surrēctum I rise, get up
sus-cipiō, -ere, -cēpī, -ceptum I undertake
suspīci-ō, -ōnis, f. suspicion
su-spiciō, -ere, -spexī, -spectum I suspect
suspicor (1) I suspect
sus-tineō, -ēre, -tinuī I sustain, bear up, support
sustulī – see tollō

taceō (2) I am silent
tacit-us, -a, -um silent
taedet (mē) I am tired of
tāl-is, -e such
tam so
tamen but, however
tamquam (sī) as if
tandem at length
tangō, -ere, tetigī, tāctum I touch
tantum only
tant-us, -a, -um so great
tard-us, -a, -um slow
tēct-um, -ī, n. roof, house
tegō, -ere, tēxī, tēctum I cover
tell-ūs, -ūris, f. the earth, land
tēl-um, -ī, n. missile
tempest-ās, -ātis, f. storm, weather
templ-um, -ī, n. temple
temptō (1) I try, attempt
temp-us, -oris, n. time
tenebr-ae, -ārum, f. pl. darkness
teneō, -ēre, tenuī, tentum I hold
tener, tenera, tenerum tender
terg-um, -ī, n. back
  ā tergō from behind

terr-a, -ae, f. earth; land
terreō (2) I terrify
terror, -is, m. terror
testimōni-um, -ī, n. evidence
test-is, -is, c. witness
timeō (2) I fear
timidit-ās, -ātis, f. fear, nervousness
timid-us, -a, -um timid
timor, -is, m. fear
tog-a, -ae, f. toga
tollō, -ere, sustulī, sublātum I raise, remove,
  destroy
torqueō, -ēre, torsī, tortum I twist, turn
tot (indecl.) so many
tōt-us, -a, -um whole
trā-dō, -ere, -didī, -ditum I hand over, and down
trāgul-a, -ae, f. javelin
trahō, -ere, trāxī, tractum I draw, drag
trā-iciō, -ere, -iēcī, -iectum I throw across, send
  across; I cross
trāns + acc. across
trāns-eō, -īre, -iī, -itum I cross
trāns-gredior, -ī, -gressus I cross
trānsit-us, -ūs, m. crossing
trānsportō (1) I transport
tribūnus plēbis tribune of the people
trīclīni-um, -ī, n. dining room
trīstis, -e sad
triumphō (1) I triumph
triumph-us, -ī, m. triumph
tueor (2) I gaze at, observe; I protect
tum then
tumult-us, -ūs, m. commotion, uproar; rebellion
tumul-us, -ī, m. tomb; hillock
tunc then
turb-a, -ae, f. crowd
turp-is, -e disgraceful
turr-is, -is, f. tower
tūt-us, -a, -um safe

ūber, -is fertile
ubi when
ubi? where?
ubīque everywhere
ulcīscor, -ī, ultus I avenge
ūllus, -a, -um any
ulteri-or, -us further
ultim-us, -a, -um furthest, last
ultrā + acc. beyond
umbra, -ae, f. shadow
umer-us, -ī, m. shoulder
umquam ever
ūnā together
und-a, -ae, f. wave
unde? whence?
undique from all sides
unquam ever
urbān-us, -a, -um of the city,
urbs, urbis, f. city
urgeō, -ēre I push, press

usquam anywhere
usque continually
usque ad right up to
ūs-us, -ūs, *m.* use
ut prīmum as soon as
uter, utra, utrum? which of two?
uterque, utraque, utrumque each of two
utī = ut
ūtil-is, -e useful
ūtor, -ī, ūsus + abl. I use
utrum . . . an . . . whether . . . or . . .
uxor, -is, *f.* wife

vacu-us, -a, -um empty
vagor (1) I wander
valdē very
valeō (2) I am well; I am strong
valē farewell
valētūd-ō, -inis, *f.* health
valid-us, -a, -um strong
vallēs, vallis, *f.* valley
vāll-um, -ī, *n.* rampart
vari-us, -a, -um various, different
-ve or
vehemēn-s, -tis violent
vehō, -ere, vēxī, vectum I carry
velut like; just as (if)
vendō, -ere, vendidī, venditum I sell
venēn-um, -ī, *n.* poison
veni-a, -ae, *f.* pardon
veniō, -īre, vēnī, ventum I come
vēnor (1) I hunt
vent-er, -ris, *m.* belly, stomach
vent-us, -ī, *m.* wind
vēr, vēris, *n.* spring
verb-um, -ī, *n.* word
vereor (2) I fear
vērō in fact, certainly; but
versor (1) I take part in
vers-us, -ūs, *m.* verse
vertō, -ere, vertī, versum I turn
vēr-us, -a, -um true
vesper, -is, *m.* evening

vest-er, -ra, -rum your
vestīgi-um, -ī, *n.* footprint, track, trace
vestiō (4) I clothe
vest-is, -is, *f.* clothing
vi-a, -ae, *f.* road, way
vetō, -āre, vetuī, vetitum I forbid
vet-us, -eris old
vexō (1) I harass, annoy
vīcīn-us, -a, -um neighbouring
victor, -is, *m.* conqueror
victōri-a, -ae, *f.* victory
videō, -ēre, vīdī, vīsum I see
videor, -ērī, vīsus I seem, appear
vigili-a, -ae, *f.* watch
vigilō (1) I am awake
vīll-a, -ae, *f.* country house
vinciō, -īre, vīnxī, vīnctum I bind
vincō, -ere, vīcī, victum I conquer
vincul-um, -ī, *n.* chain
vīn-um, -ī, *n.* wine
vir, -ī, *m.* man
virt-ūs, -ūtis, *f.* manliness, courage
vīs, vim, vī, *f.* force
vīrēs, -ium, *f.pl.* strength
vīsō, -ere, vīsī I visit, go to see
vīt-a, -ae, *f.* life
viti-um, -ī, *n.* fault, vice
vītō (1) I avoid
vituperō (1) I abuse
vīvō, -ere, vīxī, vīctum I live
vīv-us, -a, -um alive
vix scarcely
vocō (1) I call
volō (1) I fly
volō, velle, voluī I wish, am willing
volunt-ās, -ātis, *f.* will, good will
volupt-ās, -ātis, *f.* pleasure
volvō, -ere, volvī, volūtum I roll
vōt-um, -ī, *n.* prayer, vow
vōx, vōcis, *f.* voice
vulg-us, -ī, *m.* the people
vulnerō (1) I wound
vuln-us, -eris, *n.* wound
vult-us, -ūs, *m.* face, expression

ENGLISH-LATIN VOCABULARY

able, I am **possum, posse, potuī**
about (= concerning) **dē** + abl.
accuse, I **accūsō** (1)
accustomed, I am **soleō, -ēre, solitus**
across **trāns** + acc.
advance, I **prō-gredior, -ī, -gressus**
advise, I **moneō** (2)
Aeneas **Aenē-ās, -ae**
afraid, I am **timeō** (2)
after (conjunction) **postquam**
again **iterum**
against one's will **invīt-us, -a, -um**
all **omn-is, -e**
allowed, I am **mihi licet** (2)
almost **paene**
already **iam**
altar **ār-a, -ae,** *f.*
although **quamquam**
always **semper**
ambassador **lēgāt-us, -ī,** *m.*
among **inter** + acc.
angered **īrāt-us, -a, -um**
animal **anim-al, -ālis,** *n.*
appear **appāreō** (2)
approach, I **appropinquō** (1) + dat.
arise, I **orior, orīrī, ortus**
arms **arm-a, -ōrum,** *n.pl.*
army **exercit-us, -ūs,** *m.*
arrest, I **com-prehendō, -ere, -prehendī, -prehēnsum**
arrival **advent-us, -ūs,** *m.*
arrive, I **ad-veniō, -īre, -vēnī, -ventum**
art **ars, artis,** *f.*
ashamed, I am **mē pudet**
as . . . as . . . **tam . . . quam . . .**
ask, I (= invite) **vocō** (1)
ask, I (a question) **rogō** (1)
ask for, I **rogō** (1)
astonished **attonit-us, -a, -um**
at last **tandem**
at length **tandem**
attack, I **oppugnō** (1)

bad **mal-us, -a, -um**
bank **rīp-a, -ae,** f.
battle **proeli-um, -ī,** n.
beautiful **pulch-er, -ra, -rum**
because **quod**
become, I **fīō, fierī, factus**
before (preposition) **ante** + acc.
before (conjunction) **antequam**
before (adverb) **anteā**
begin I **in-eō, -īre, -iī, -itum; coepī, coepisse**
believe, I **crē-dō, -ere, -didī, -ditum** + dat.
best **optim-us, -a, -um**
betray, I **prō-dō, -ere, -didī, -ditum**

better **mel-ior, -ius**
birth (adjective) **nātāl-is, -e**
body **corp-us, -oris,** *n.*
borders **fīn-ēs, -ium,** *m.pl.*
both . . . and . . . **et . . . et . . .**
brave **fort-is, -e**
    bravely **fortiter**
bring, I **dūcō, -ere, dūxī, ductum**
          ferō, ferre, tulī, lātum
Britain **Britanni-a, -ae,** *f.*
Briton **Britann-us, -ī,** *m.*
brother **frāt-er, -ris,** *m.*
build, I **aedificō** (1)
build, I (of a ship) **cōn-struō, -ere, -strūxī, -strūctum**
but **sed**
buy, I **emō, emere, ēmī, ēmptum**
by **ā/ab** + abl.
by (of place) **ad** + acc.

call, I **appellō** (1), **vocō** (1)
calm **tranquill-us, -a, -um**
camp **castr-a, -ōrum,** *n.pl.*
can, I **possum, posse, potuī**
careful **dīlig-ēns, -entis**
Carthaginian **Poen-us, -ī,** *m.*
catch, I **capiō, capere, cēpī, captum**
cause **caus-a, -ae,** *f.*
cavalry **equit-ēs, -um,** *m.pl.*
cave **spēlunc-a, -ae,** *f.*
cease, I **dē-sinō, -ere, -siī, -situm**
certain, a **quīdam, quaedam, quoddam**
certainly **certē**
chariot **curr-us, -ūs,** *m.*
chief **prīn-ceps, -cipis,** *m.*
city **urbs, urbis,** *f.*
clothing **vest-is, -is,** *f.*
coast **ōr-a, -ae,** *f.*
collapse, I **con-cidō, -ere, -cidī**
collect, I **col-ligō, -ere, -lēgī, -lēctum**
column **agm-en, -inis,** *n.*
come, I **veniō, -īre, vēnī, ventum**
conceal **cēlō** (1)
condition **condici-ō, -ōnis,** *f.*
conspirator **coniūrāt-us, -ī,** *m.*
consul **cōns-ul, -ulis,** *m.*
contest **certām-en, -inis,** *n.*
corn **frūment-um, -ī,** *n.*
country (= land) **terr-a, -ae,** *f.*
courage **virt-ūs, -ūtis,** *f.*
coward **ignāv-us, -ī,** *m.*
cross (over), I **trāns-eō, -īre, -iī, -itum**
crowd **turb-a, -ae,** *f.*
cruel **crūdēl-is, -e**

danger **perīcul-um, -ī,** *n.*

dare, I **audeō, -ēre, ausus**
daughter **fīli-a, -ae,** *f.*
dawn **prīma lūx, prīmae lūcis**
day **di-ēs, -ēī,** *m.*
dead **mortu-us, -a, -um**
dear **cār-us, -a, -um**
death **mors, mortis,** *f.*
decide, I **cōn-stituō, -ere, -stituī, -stitūtum**
deed **fact-um, -ī,** *n.*
deep **alt-us, -a, -um**
defeat, I **superō** (1); **vincō, -ere, vīcī, victum**
defend, I **dē-fendō, -ere, fendī, -fēnsum**
delay, I **moror** (1)
descend, I **dē-scendō, -ere, -scendī, -scēnsum**
despair, I **dēspērō** (1)
destroy, I **dē-leō, -ēre, -lēvī, -lētum**
Dido **Dīd-ō, -ōnis**
die, I **morior, -ī, mortuus**
dine, I **cēnō** (1)
dinner **cēn-a, -ae,** *f.*
discover, I **reperiō, -īre, repperī, repertum**
disembark, I **ē nāve/nāvibus ēgredior, -ī, ēgressus**
do, I **faciō, -ere, fēcī, factum; ef-ficiō, -ere, -fēcī,
   -fectum**
door **iānu-a, -ae,** *f.*
draw up, I **īn-struō, -ere, -strūxī, -strūctum**
drink, I **bibō, -ere, bibī**
duty **offici-um, -ī,** *n.*
dwell in, I **habitō** (1)

eager **avid-us, -a, -um**
easy **facil-is, -e**
embrace, I **am-plector, -ī, -plexus**
emperor **imperāt-or, -ōris,** *m.*
encourage, I **hortor** (1)
enemy **host-is, -is,** *c.;* **host-ēs, -ium,** *m.pl.*
enough **satis**
especially **praecipuē**
eternal **aetern-us, -a, -um**
ever **umquam**
ever (= always) **semper**
everywhere **ubīque**
evil **prāv-us, -a, -um**
expense **sūmpt-us, -ūs,** *m*
eye **ocul-us, -ī,** *m.*

fall, I **cadō, -ere, cecidī**
family **gen-us, -eris,** *n.*
famous **clār-us, -a, -um**
far, I am **absum, abesse, āfuī**
father **pat-er, -ris,** *m.*
fear, I **timeō** (2)
fierce **ācer, ācris, ācre**
fight, I (intrans.) **pugnō** (1)
fight with, I **pugnō** (1) **cum** + abl.
finally **tandem**
find, I **in-veniō, -īre, -vēnī, -ventum**
find out, I **cog-nōscō, -ere, -nōvī, -nitum**
first **prīm-us, -a, -um**
   first (adverb) **prīmum**

at first **prīmō**
flee, I **fugiō, -ere, fūgī, fugitum**
fleet **class-is, -is,** *f.*
follow, I **sequor, -ī, secūtus**
for **nam**
forces **cōpi-ae, -ārum,** *f.pl.*
forget, I **ob-līvīscor, -ī, -lītus** + gen.
forgive, I **ig-nōscō, -ere, -nōvī** + dat.
fortunate **fortūnāt-us, -a, -um**
found, I **con-dō, -ere, -didī, -ditum**
fountain **fōns, fontis,** *m.*
four **quattuor**
friend **amīc-us, -ī,** *m.*
from **ē/ex** + abl.

Gaul **Gallia, -ae,** *f.*
Gaul, a **Gall-us, -ī,** *m.*
gather I **col-ligō, -ere, -lēgī, -lēctum**
general **dux, ducis,** *m.*
ghost **imāg-ō, -inis,** *f.*
girl **puell-a, -ae,** *f.*
give, I **dō, dare, dedī, datum**
glory **glōri-a, -ae,** *f.*
go, I **eō, īre, iī/īvī**
go away, I **ab-eō, -īre, -iī**
go down, I **dē-scendō, -ere, -scendī, -scēnsum**
go forward, I **prō-gredior, -gredī, -gressus**
go out, I **ex-eō, -īre, -iī**
go to, I **ad-eō, -īre, -iī**
god **de-us, -ī,** *m.*
goddess **de-a, -ae,** *f.*
gold **aur-um, -ī,** *n.*
good **bon-us, -a, -um**
govern, I **administrō** (1)
greatly **magnopere**
greet, I **salūtō** (1)
greetings **sal-ūs, -ūtis,** *f.*
ground **sol-um, -ī,** *n.*
guard, I **custōdiō** (4)

handsome **fōrmōs-us, -a, -um**
Hannibal **Hannibal, -is**
Hanno **Hann-ō, -ōnis**
happens, it **accidit**
happy **laet-us, -a, -um**
harm, I **noceō** (2) + dat.
hate, I **ōdī, ōdisse**
have, I **habeō** (2)
hard (adverb) **dīligenter**
health **valētūd-ō, -inis,** *f.*
hear, I **audiō** (4)
help, I **iuvō, -āre, iūvī, iūtum**
help **auxili-um, -ī,** *n.*
here, from **hinc**
hill **coll-is, -is,** *m.*
hold, I **teneō, -ēre, tenuī, tentum**
home **dom-us, -ūs,** *f.*
   at home **domī**
hostage **obs-es, -idis,** *c.*
hope, I, **spērō** (1)

hour **hōr-a, -ae,** *f.*
house **dom-us, -ūs,** *f.*
how? **quōmodo?**
how are you? **quid agis?**
however **tamen** (2nd word of sentence)
huge **ing-ēns, -entis**
hurry, I **festīnō** (1)
husband **marīt-us, -ī,** *m.*

if **sī**
into **in** + acc.
invade, I **in-vādō, -ere, -vāsī, -vāsum**
indignant, I am **indignor** (1)
Italy **Itali-a, -ae,** *f.*

journey **iter, itineris,** *n.*
join, I **iungō, -ere, iūnxī, iūnctum**
join battle, I **proelium com-mittō, -ere, -mīsī,
   -missum**
joy **gaudi-um, -ī,** *n.*
jump, I **saliō, -īre, saluī**
Juno **Iūn-ō, -ōnis**
Jupiter **Iuppiter, Iovis**
justice **iūstiti-a, -ae,** *f.*
justice, administer, I **iūs dīcō**

kill, I **oc-cīdō, -ere, -cīdī, -cīsum**
king **rēx, rēgis,** *m.*
know, I **sciō** (4)
know, I do not **nesciō** (4)

land, I **ē-gredior, -ī, -gressus**
land **terr-a, -ae,** *f.*
large **magn-us, -a, -um**
later (adverb) **posteā**
law **lēx, lēgis,** *f.*
lead, I **dūcō, -ere, dūxī, ductum**
leader **dux, ducis,** *m.*
learn, I (= find out) **cog-nōscō, -ere, -nōvī, -nitum**
leave, I (= go away) **dis-cēdō, -ere, -cessī, -cessum**
leave, I (= go away from) **re-linquō, -ere, -līquī,
   -lictum**
legion **legi-ō, -ōnis,** *f.*
leisure, at **ōtiōs-us, -a, -um**
letter **epistul-a, -ae,** *f.*
life **vīt-a, -ae,** *f.*; **aet-ās, -ātis,** *f.*
lion **le-ō, -ōnis,** *m.*
listen to, I **audiō** (4)
live, I **vīvō, -ere, vīxī, vīctum**
live in, I **habitō** (1)
love, I **amō,** (1)
love **amor, -is,** *m.*
lover **amātor, -is,** *m.*

make, I **faciō, -ere, fēcī, factum**
man **vir, -ī,** *m.*
may (I may) **mihi licet**
many **mult-ī, -ae, -a**
march, I **con-tendō, -ere, -tendī, -tentum**
meet, I **oc-currō, -ere, -currī, -cursum** + dat.
meet, I (= gather together) **con-veniō, -īre, -vēnī,
   -ventum**

merchant **mercātor, -is,** *m.*
message **nūnti-us, -ī,** *m.*
messenger **nūnti-us, -ī,** *m.*
money **argent-um, -ī,** *n.*
mother **māt-er, -ris,** *f.*
month **mēns-is, -is,** *m.*
mount, I **as-cendō, -ere, -scendī, -scēnsum**
mountain **mōns, montis,** *m.*
mind **anim-us, -ī,** *m.*
must, I **dēbeō** (2)

native **indigen-a, -ae,** *m.*
near **prope** + acc.
new **nov-us, -a, -um**
next **poster-us, -a, -um**
next to **iuxtā** + acc.
night **nox, noctis,** *f.*
at night **noctū**
nobody **nēm-ō, -inis**
nothing **nihil**

obey, I **pāreō** (2) + dat.
often **saepe**
old **antīqu-us, -a, -um, nāt-us, -a, -um**
old man **senex, senis,** *m.*
on **in** + abl.
once **ōlim**
once (as opposed to *twice*) **semel**
once, at **statim**
one, the . . . the other . . . **alter, -a, -um . . . alter,
   -a, -um . . .**
onslaught **impet-us, -ūs,** *m.*
opportunity **occāsi-ō, -ōnis,** *f.*
order, I **imperō** (1) + dat.; **iubeō, -ēre, iussī,
   iussum**
order **imperi-um, -ī,** *n.*
other **ali-us, -a, -ud**
other (the other) **alter, altera, alterum**
oppress, I **opprim-ō, -ere, -pressī, -pressum**
outstanding **praest-āns, -antis**
over **trāns** + acc.
overcome, I **superō** (1)
overthrow, I **ē-vertō, -ere, -vertī, -versum**
Ovid **Ovidi-us, -ī**

part **pars, partis,** *f.*
persuade, I **per-suādeō, -ēre, -suāsī, -suāsum**
   + dat.
parrot **psittac-us, -ī,** *m.*
Parthian **Parth-us, -ī,** *m.*
parent **par-ēns, -entis,** *c.*
peace **pāx, pācis,** *f.*
people **popul-us, -ī,** *m.*
people (plural of *person*) **homin-ēs, -um,** *m. pl.*
perform, I **per-ficiō, -ere, -fēcī, -fectum**
place **loc-us, -ī,** *m.*
place, I **locō** (1)
please, I **placeō** (2) + dat.
poet **poēt-a, -ae,** *m.*
poetry **carmin-a, -um,** *n. pl.*
Pompey **Pompēi-us, -ī,** *m.*

port **port-us, -ūs,** *m.*
praise, I **laudō** (1)
prefer, I **mālō, māllé, māluī**
present, I am **ad-sum, -esse, -fuī**
pretty **pulcher, pulchra, pulchrum**
prepare, I **parō** (1)
province **prōvinci-a, -ae,** *f.*
purple **purpure-us, -a, -um**
pursue, I **per-sequor, -ī, -secūtus**
proud **superb-us, -a, -um**
prayers **prec-ēs, -um,** *f. pl.*
promise, I **prō-mittō, -ere, -mīsī, -missum**
punish, I **pūniō** (4)
propose, I **prō-pōnō, -ere, -posuī, -positum**
pyre **pyr-a, -ae,** *f.*

queen **rēgīn-a, -ae,** *f.*
quickly **celeriter**

race (= horse race) **certām-en, -inis,** *n.*
rampart **vāll-um, -ī,** *n.*
realize, I **sentiō, -īre, sēnsī, sēnsum**
reach, I **per-veniō, -īre, -vēnī, -ventum ad**
rear, from the **ā tergō**
recall, I **revocō** (1)
receive, I **ac-cipiō, -ere, -cēpī, -ceptum**
recite, I **recitō** (1)
reject, I **rē-iciō, -ere, -iēcī, -iectum**
remain, I **maneō, -ēre, mānsī, mānsum**
reply, I **re-spondeō, -ēre, -spondī, -spōnsum**
republic **rēspūblica, reīpūblicae,** *f.*
resist, I **re-sistō, -ere, -stitī** + dat.
respectable **honest-us, -a, -um**
rest, I **quiē-scō, -ere, -vī, -tum**
return, I **red-eō, -īre, -iī, -itum**
return **redit-us, -ūs,** *m.*
retire, I **re-cēdō, -ere, -cessī, -cessum**
river **flūm-en, -inis,** *n.*
Roman **Rōmān-us, -a, -um**
Rome **Rōm-a, -ae,** *f.*
  at Rome **Rōmae**
rouse, I **excitō** (1)
run, I **currō, -ere, cucurrī, cursum**

sad **trīst-is, e**
safe **tūt-us, -a, -um**
sail, I **nāvigō** (1)
savage **saev-us, -a, -um**
save, I **servō** (1)
say, I **dīcō, -ere, dīxī, dictum**
scarcely **vix**
sea **mar-e, -is,** *n.*
see, I **videō, -ēre, vīdī, vīsum**
seem, I **videor, -ērī, vīsus**
seize, I **rap-iō, -ere, -uī, -tum,**
seldom **rārō**
self **ips-e, -a, -um**
senate **senāt-us, -ūs,** *m.*
send, I **mittō, -ere, mīsī, missum**
send forth, I **ē-mittō, -ere, -mīsī, -missum**
set (of the sun) **oc-cidō, -ere, -cidī**

set out, I **pro-ficīscor, -ī, -fectus**
shade **umbr-a, -ae,** *f.*
shadows **tenebr-ae, -ārum,** *f. pl.*
ship **nāv-is, -is,** *f.*
shore **līt-us, -oris,** *n.*
short of, I am **careō** (2) + abl.
shut (in), I **in-clūdō, -ere, -clūsī, -clūsum**
shout, I **clāmō** (1)
show, I **ostendō, -ere, -tendī, -tentum**
show (= a public show) **mūn-us, -eris,** *n.*
side (on both sides, etc.) **pars, partis,** *f.*
silent **tacit-us, -a, -um**
since (= after) **post** + acc.
sister **soror, -is,** *f.*
sit, I **sedeō, -ēre, sēdī**
slave **serv-us, -ī,** *m.*
smile at, I **sub-rīdeō, -ēre, -rīsī** + dat.
smoke **fūm-us, -ī,** *m.*
snow **nix, nivis,** *f.*
so (with adjectives) **tam**
so many **tot** (indecl. adj.)
so much **adeō**
someone **aliquis, aliquid**
son **fīli-us, ī,** *m.*
soon **mox**
soul **anim-a, -ae,** *f.*
speak, I **loquor, -ī, locūtus**
spend, I (of time) **agō, -ere, ēgī, āctum**
speed **celerit-ās, -ātis,** *f.*
spring **vēr, vēris,** *n.*
staff **cohors, cohortis,** *f.*
standard **sign-um, -ī,** *n.*
stay, I **maneō, -ēre, mānsī, mānsum**
stop, I **cōn-sistō, -ere, -stitī**
storm **tempest-ās, -ātis,** *f.*
study, I **studeō** (2) + dat.
such **tāl-is, -e**
such = so great **tant-us, -a, -um**
suddenly **subitō**
suffer, I **patior, -ī, passus**
suitable **idōne-us, -a, -um**
summer **aet-ās, -ātis,** *f.*
sun **sōl, -is,** *m.*
surely not? **num?**
surround, I **circum-veniō, -īre, -vēnī, -ventum**
swear, I **iūrō** (1)
swim, I **natō** (1)

take, I **capiō, -ere, cēpī, captum**
take back, I **reportō** (1)
talk, I **loquor, -ī, locūtus**
tell, I (= inform) **certiōrem faciō**
tell, I (= order) **imperō** (1) + dat.
territory **fīn-ēs, -ium,** *m. pl.*
Thames **Tames-is, -is,** *m.*
than **quam**
then **deinde**
there **ibi**
there (= to there) **eō**
thing **rēs, reī,** *f.*

think, I **putō** (1)
threaten, I **minor** (1) + dat.
through **per** + acc.
thus **ita**
time **temp-us, -oris**, *n.*
tongue **lingu-a, -ae**, *f.*
too **et; quoque**
town **oppid-um, -ī**, *n.*
transport, I **trānsportō** (1)
travel, I **iter faciō, -ere, -fēcī, -factum**
tree **arb-or, -oris**, *f.*
Trojan **Trōiān-us, -a, -um**
troubles **mal-a, -ōrum**, *n. pl.*
Troy **Trōi-a, -ae**, *f.*
trust, I **cōn-fīdō, -ere, -fīsus** + dat.
try, I **cōnor** (1)
turn, I **mē vertō, -ere, vertī**
twice **bis**

under **sub** + abl.
underworld **īnfer-ī, -ōrum**, *m. pl.*
unhappy **īnfēl-īx, -īcis**
unless **nisi**
unwilling **invīt-us, -a, -um**
upset **commōt-us, -a, -um**
use, I **ūtor, ūtī, ūsus** + abl.

valley **vall-ēs, -is**, *f.*
value, I **aestimō** (1)
verse **vers-us, -ūs**, *m.*
very much, greatly **magnopere**
vex, I **vexō** (1)

wait, I **man-eō, -ēre, -sī, -sum**
wait for, I **exspectō** (1)
walk, I **ambulō** (1)
wall **pari-ēs, -ētis**, *m.*
walls **moeni-a, -um**, *n. pl.*
want, I **volō, velle, voluī; cupiō, -ere, cupīvī,
    cupītum**

want, I do not **nōlō, nōlle, nōluī**
war **bell-um, -ī**, *n.*
warn, I **admoneō** (2)
water **aqu-a, -ae**, *f.*
wave **und-a, -ae**, *f.*
weak **īnfirm-us, -a, -um**
well **bene**
what? **quid?**
where? **ubi?**
where from? **unde?**
whether . . . or . . . **utrum . . . an . . .**
while **dum**
white **candid-us, -a, -um**
who? **quis? quid?**
whole **tōt-us, -a, -um**
why? **cūr?**
wide **lāt-us, -a, -um**
wife **uxor, -is**, *f.*
wild animal **fer-a, -ae**, *f.*
win, I **vincō, -ere, vīcī, victum**
wind **vent-us, -ī**, *m.*
wish, I **volō, velle, voluī; optō** (1)
winter **hiem-s, -is**, *f.*
withstand, I **re-sistō, -ere, -stitī; sus-tineō, -ēre,
    -tinuī, -tentum**
wonder, I (= want to know) **scīre volō**
wonderful **mīr-us, -a, -um**
wood **silv-a, -ae**, *f.*
word **verb-um, -ī**, *n.*
write, I **scrībō, -ere, scrīpsī, scrīptum**
wrong **iniūri-a, -ae**, *f.*
wrong, I am **errō** (1)
worried **ānxi-us, -a, -um**
wound, I **vulnerō** (1)

year **ann-us, -ī**, *m.*
yesterday **herī**
young man **iuven-is, -is**, *m.*
your (singular) **tu-us, -a, -um**

INDEX OF GRAMMAR

Most basic grammar and much basic syntax is covered in Parts I and II (see indices of grammar for these Parts). The brief summary of Latin syntax (pages 228–230), which covers all the syntax introduced into this course, may be useful for revision.